THE STATE OF
CHINA
ATLAS

Stephanie Hemelryk Donald
is Professor of Communication and Culture,
and Director of Transforming Cultures
at the Centre for Communication and Culture,
University of Technology, Sydney, Australia.
She is author of *Public Spaces, Public Secrets: Cinema
and Civility in China* and *Little Friends: Children's Film and
Media Culture in China*. She also co-authored *The
Penguin Atlas of Media and Information*, and co-edited
Media in China: Consumption, Content and Crisis. She is
Series Editor of the RoutledgeCurzon series "Media,
Society and Change in Asia-Pacific."

Robert Benewick
is Research Professor in the Department of
International Relations, University of Sussex,
and a Visiting Fellow in the Centre for the Study of
Democracy, Westminster University, London.
His recent publications include *Asian Politics in
Development* with Marc Blecher and Sarah Cook (eds),
Contemporary China with Paul Wingrove,
and papers in *China Information* and
Community Construction in China.

THE STATE OF
CHINA
ATLAS

Stephanie Hemelryk Donald

and

Robert Benewick

UNIVERSITY OF CALIFORNIA PRESS

Berkeley   Los Angeles   London

This second edition first published by
University of California Press
Berkeley and Los Angeles, California
University of California Press Ltd
London, England
in 2005

10 9 8 7 6 5 4 3 2 1

ISBN 0-520-24627-6

Cataloging-in-publication data is on file with
the Library of Congress

Produced for University of California Press by
Myriad Editions
6–7 Old Steine, Brighton BN1 1EJ, UK
Email: info@MyriadEditions.com
Web: www.MyriadEditions.com

Edited and coordinated for Myriad Editions by
Jannet King, Candida Lacey and Sadie Mayne
Design and graphics by Isabelle Lewis and Corinne Pearlman
Cartography by Isabelle Lewis

Printed and bound in Hong Kong through Phoenix Offset
under the supervision of Bob Cassels,
The Hanway Press, London

CONTENTS 中

Foreword by Professor Tony Saich   7
Introduction   9

Part One CHINA IN THE WORLD   12
1   Trade   14
2   Investing in China   16
3   Military Power   18
4   International Relations   20

Part Two CHINA'S PEOPLE   22
5   Population   24
6   Urbanization   26
7   The Gender Gap   28
8   Minority Nationalities   30

Part Three THE ECONOMY   32
9   Entrepreneurial China   34
10   Equality and Inequality   36
11   Employment   38
12   Agriculture   40
13   Industry   42
14   Services   44
15   Tourism   46
16   Traffic   48
17   Energy   50

Part Four THE PARTY-STATE   52
18   The Chinese Communist Party   54
19   Central Government   56
20   The People's Liberation Army   58
21   The Center and the Provinces   60
22   Rule According to Law   62
23   State *v* Citizens   64

Part Five LIVING IN CHINA   66
24   Households   68
25   Food   70
26   Education   72
27   Welfare   74
28   Health   76
29   Tobacco   78
30   Telecommunications   80
31   Media   82
32   Religion   84

Part Six THE ENVIRONMENT   86
33   Air Pollution   88
34   Water Resources   90
35   Water Health   92

Part Seven TABLES   94
China in the World   96
China: Population   98
China: Economy   100
China: Living and Lifestyle   102
China: Natural Resources   104

Commentaries   108
Select Bibliography   126
Index   127

In memory of

Anne Benewick

This latest edition of *The State of China Atlas* is a welcome resource for anyone who wants to understand the complexity of a rapidly changing China. It is no exaggeration to say that the future of us all is bound up in one way or another with what is happening and will happen in China. Thus, it is important for us to be able to make sense of this endlessly fascinating country. But how do we do that? As the saying goes, a picture speaks a thousand words, and this is certainly true of this wonderful volume. This Atlas provides one of the best entry points to understanding China in all its variety. The introductions are concise and precise, while the maps are well chosen and visually striking. They provide all the basic information and more that anyone wanting to get to grips with China needs. Even for those who have been studying China for many years, the maps present new insights, and the volume is a key reference work. For students, the Atlas is a dynamic and exciting way to bring China alive. It contains so much useful information on every topic crucial to China's development. The authors have even found a way to capture the party-state in visual form – no mean achievement.

What is revealed in these pages is a China where multiple realities operate beneath the facade of a unitary nation-state. Not only does the terrain range from the huge oceans to the east, the massive plateau to the west and the surrounding mountains that have helped China retain a certain insularity, but also the peoples of China, the climate, its industry and its agriculture show tremendous diversity. This diversity defies the easy descriptions and classifications that appear in the popular press, which cloud our understanding of China.

The State of China Atlas shows how, since 1949, the Chinese Communist Party (CCP) has placed its own stamp on this varied terrain. The CCP's vision of a modern state, and its policies of industrialization, have had a marked impact on the physical structure of towns and countryside as well as on people's lives. On taking power, the CCP laid out a vision of the future that was inspired by the Soviet Union. A modern China would be one that was urban and industrial, with production socialized. The private sphere was to be destroyed and the rural sector was to be placed in the service of the industrial push.

This produced an ugly urban landscape, and even historic cities such as Beijing had much of the grace and charm ripped out them by Soviet-inspired planners. Old lanes and alleyways were plowed under to accommodate the new, wider roads and workplace apartment blocks. The industrial push made smokestack factories a familiar part of many cities, with little or no idea of zoning and protection of green areas. The countryside was also transformed, culminating in the commune movement in the late-1950s. This brought a greater uniformity to the visual impression of rural life than one would expect, given the varied topography. Campaigns to boost grain production led to mountain slopes being cleared of trees, and good animal grazing land being plowed under to plant the seeds that would help each locality reach their inflated targets. The heyday of communization was accompanied by rural industrialization that led to more forests being ripped up to produce steel in "backyard furnaces" – much of which was useless.

Despite such attempts to produce a dull conformity, China remained full of contradictions and variety, and these have been allowed to blossom again since economic reforms were promoted with such zeal from the late-1970s on. As the Atlas shows, these reforms have touched on every aspect of Chinese life. They have changed the physical look of both rural and urban China while binding the two closer together than in the Mao years. Cities are less homogeneous than before, and the drab Stalinesque town centers have been transformed by the rise of gleaming, glass-fronted skyscrapers housing luxury offices, shopping malls and the ubiquitous McDonald's. These buildings, and designer brands such as Gucci, are the new symbols of modernization, and much of the old architecture that survived the Maoist blitz has been bulldozed out of the way. In fact, in the "go-go" 1990s more of historical Beijing was lost to the real-estate developers than in any other decade of the century (decades that included civil war, Japanese invasion and the Cultural Revolution). Communities have been broken up and scattered in the name of modernity. Much of the new building that is not commercial is to reify state and party power, with many, new, gleaming, marble-decked buildings constructed to house the local party, government and judicial organs of the state.

Beneath the high rises, the Chinese streets are home to a much more diverse life. The markets, restaurants and discos are signs of the new entrepreneurship, or official organizations moonlighting to make a bit of extra money. The restaurants are filled with the beneficiaries of reform: the private entrepreneurs, those involved in the new economy, the managerial elites, the politically well connected and the foreigners. There are also the millions of migrants who have poured into the cities from the countryside to build the new urban "nirvanas". They staff the construction sites, work as waiters, shop assistants, masseuses, and in the less acceptable areas of vice and prostitution. Not all have been blessed by this tremendous boom and economic growth. There are the new urban poor who have been laid-off from the old state-owned factories, or who have no dependants to look after them in old age. As the Atlas shows, the cost in terms of pollution is high, as it is for traffic congestion and death from traffic accidents.

The countryside has also changed, with the collectives broken up and farming responsibility placed back with the households. This has allowed more diversification in agricultural production, and permitted millions to leave the land to find more remunerative work in the small township factories, the construction sites of big cities, or in the joint-venture factories of South China. Those who move are the young, the fit and the adventurous. Those who remain behind to tend to the farm and household chores are the elderly, married women, the children, and the sick.

These changes not only impact on China but also have worldwide ramifications. How to deal with the rising economic power of China has become a cottage industry for publishing. China is now the largest recipient of foreign direct investment in the world, most multinational corporations must have a China strategy and many countries are trying to align their own production strategies to meet China's developmental needs. China's economic needs, with the incumbent energy increases, are altering global markets for natural resources, and prices will be increasingly determined by projections of China's needs. There will be consequences that move beyond the purely economic. Already, Japan is badly affected by industrial emissions from China, which is also a major producer of greenhouse gases. Decisions made in China can affect other nations in unexpected ways. For example, the ban by the central government on logging in southwest China is eminently sensible. However, it has not stopped China's desire for raw wood for its domestic and export markets. This is leading to an increase in logging, not only in surrounding countries such as Laos but even in those farther away, such as Brazil.

The State of China Atlas provides a good starting point for trying to unravel the consequences of these changes. It is not only informative but also fun to read and look at. It is highly recommended for all those interested in the momentous changes taking place in China.

Tony Saich
John F Kennedy School of Government
Harvard University
January 2005

INTRODUCTION 中

WE INTRODUCED THE 1999 EDITION of the atlas with a selection of widely held impressions of China: exotic, esoteric, mystical, even mysterious. Although these impressions may linger in some people's minds, they have now been supplemented by the idea of China as a modernizing economic giant. The new vision of China is of an important political power entering the global community of nations on its own terms. Notwithstanding this changing image, China still remains mysterious to many people. Commerce and extended trade relations do not counteract the unknowability of profound difference.

So how is China known in the contemporary world? Whilst in the international imagination, China is bound up in extravagant symbols of development and capital, its minority peoples and most of its provinces are hardly known. Most people think of the "centralising kingdom" (*zhongguo*), as it has been conveyed through classical art, Tang poetry, through revolutionary meetings in Tiananmen – and latterly through the massacre of 1989 – through to sparkling business districts in Shanghai, to the neon presence of multinational corporations, Coca Cola and Pepsi on Nanjing Road, to the shift from most-favored nation of Bill Clinton's America to the WTO member of the Bush era in global politics. In media reports, China is Beijing or Shanghai. It is power or money. Sometimes it is a flood, the Three Gorges Dam or another terrible coal-mining disaster. But China remains essentially inscrutable and unknowable, not because the Chinese are either of these things, but because – in getting to know China – the West must recognize the limits of its assumptions, and that challenge is too hard in our own state of chronic transition and global discomfort. Yet, this is the great value in understanding the world through the state of China.

Almost every day we read that China is among the top trading nations; that its economy is one of the world's largest; and it is one of the nations attracting the most direct foreign investment. As if this is not enough, China has become one of the largest manufacturers of automobiles, where only

a few years ago a seemingly enduring and endearing picture was of workers on bicycles dominating urban traffic. Not surprisingly, for example, China is the largest user of steel and cement in the world. Cranes soaring above a cityscape is one of the images the leaders of New China wish to project as the nation prepares to host the world at the 2008 Olympics in Beijing and the 2010 Exposition in Shanghai.

There is much to trumpet, even to celebrate, and we can marvel at China's successes. Even on an ideological level, leaders of western nations derive a certain satisfaction since many of China's economic achievements can be credited to the market-led reforms that began in 1978. This may be a dominant perspective but it is not the only one. As is the case for every nation the reality is more complex. There is no doubt that most citizens in China are better off than they ever have been. An alternative perspective, however, takes into account that although there have been impressive inroads into poverty alleviation many millions remain desperately poor; a new entrepreneurial middle class is emerging, but the income gap is widening; a welfare system is being developed, yet health care remains beyond reach of most citizens; China has the most students in higher education of any nation in the world but, for many children, access to even basic education is limited.

These are examples of the contradictions that confront China's elitist and insulated leadership. They are problems familiar to other nations, but they are exacerbated by China's 1.3 billion population. China's population size can be seen as a great resource in a globalizing economy, providing a flourishing consumer market and a bottomless pool of cheap labor to exploit. It is also a source of mounting dissatisfaction, unrest and conflict, and so it is no wonder that China's authoritarian Party-state places political and social stability alongside the market-led economy as its main priorities. The Party-state does not brook opposition nor tolerate dissent, as the crackdown on the 1989 Tiananmen protest and the Falun Gong testify. Human-rights abuses continue, despite international pressure. Even more threatening to the very fabric of the Party-state is the rampant corruption. The attempt to bring corruption under control is one of a number of reforms to the political system. Another example is the introduction of participatory, if not democratic, practices at the grass-roots and basic levels of government. Reforming the Communist Party to grant more influence to the membership is also on the agenda. Whether the pace of these reforms is enough to meet the challenges to social justice is an open question.

Economic power grants China considerable leverage in international relations, but it can also be problematic. It is too soon to measure the impact of the WTO on China's economy, although in the run-up to membership the Party-state took steps to prepare and cushion sectors of the economy. The big if-and-when question within the region and the wider international system is whether there will be a struggle for dominance between the USA and China – and whether that will play out through economics and trade or some more deadly means. The current atmosphere of perpetual war, since the war on terror was declared in 2001, is frightening and would be more so if China too were pulled into the morass.

Meanwhile, statisticians and demographers in China collect large volumes of statistical information, through which they might measure the inequities across provincial and regional boundaries, and with which social scientists can interpret the state of China on the ground. It may be unfashionable but perhaps these public servants are more heroic, and certainly more functional, in determining what needs to be done in China now, and for whom. As we point out in Part Seven of this book, there are no perfect statistics, and even the collection of data is subject to political controls, both in China and worldwide. Statistics alone do not explain why one city will thrive under WTO regulations, whilst another will ignore them because they threaten local political elites, or because local businesses need local subsidy to survive and protect employment. One city will put resources and imagination into branding itself in the international imagination, whilst another will miss (or deliberately ignore) the point of tourism and "destination management".

There are maps in this book then where, to give a truly accurate picture of the state of China, we would need to provide detailed regional specifications, county and township level case studies, and a lot of historical background. Where we cannot give this detail we have suggested readings from works of current China specialists, in economy, culture, the social sciences and political history, and we really hope that readers are inspired to follow up these suggestions. The financial pages of national newspapers are also good sources for seeing certain aspects of new China unfold before our eyes. Every deal, every bankruptcy, every corporate decision will affect someone, possibly many thousands of people, in contemporary China and beyond.

As with all books, but especially one such as this, which requires a range of knowledge and expertise, the authors are indebted to the wisdom of others. The International Studies Institute research workshops at University of Technology Sydney (UTS) have been invaluable, particularly the work and comments of David Goodman, Barbara Krug, Louise Edwards, Guo Yingjie, and Feng Chongyi. The support (in time and technology) from the Transforming Cultures Research Centre and the Faculty of Humanities and Social Sciences (also UTS) has been extremely helpful. The ongoing work of colleagues in Australia and overseas: Michael Keane, Jing Wang, Anne MacClaren, Sun Wanning, John Fitzgerald, and Jane Sayers, and the APFN is always inspirational. We also thank the Australian Research Council for its support for two China projects, which have informed the arguments of this book. Marc Blecher, Rosemary Foot, Jude Howell, Qinqzhao Hua, Judith Mackay and Yawei Liu have made important contributions to the atlas and provided much-needed advice and encouragement for the whole project, while the

University of Sussex was more than generous with the provision of facilities.

We are grateful to the following for permission to reproduce their photographs: Manfred Leiter p12; DSG Goodman pp 22, 32, 66, 107; Jude Howell p 52; Curt Carnemark/The World Bank p 86; Eve Astrid Andersson <www.eveandersson.com> p 94.

We especially wish to thank our co-workers at Myriad: Candida Lacey leads with wisdom, expertise and humanity; Corinne Pearlman's and Isabelle Lewis's talents to translate hard data into comprehensible and exciting visuals is awesome; Sadie Mayne, although new to Myriad, brought a keen critical eye to both the text and the maps; and most of all, Jannet King, who as editor, researcher, critic and cajoler deserves a large share of whatever praise we receive for the atlas and none of the blame. Our greatest debt is to the memory of Anne Benewick, the co-founder of Myriad, who inspired the original edition of the atlas.

Stephanie Hemelryk Donald
Sydney, Australia

Robert Benewick
Brighton, UK

Part One

CHINA IN THE WORLD

CHINA'S REMARKABLE ECONOMIC performance has catapulted the country onto the world stage. By the end of 2005 it could well become the world's fourth-largest economy, overtaking the UK and France. But how far this will be matched in political power and influence is yet to be seen.

There is no doubt about China's ambitions, as its breathless economic growth, military modernization, and increasingly important role in international affairs testify. China's behavior today is a far cry from its distrust of the world powers, which characterized the Maoist period. A transformation of China's domestic economic and political structures has created a market-oriented economy and opened up China to the world, while maintaining an authoritarian political system.

Following the convulsions of the Cultural Revolution between 1966 and 1976, China's leader, Deng Xiaoping, and his reformist colleagues had two main priorities on their accession to power in 1978. The first was to establish and secure their legitimacy to rule. They achieved this by moving away from a command economy, with its shared deprivations, towards a market economy in order to gain popular support by raising the standard of living. There is little doubt that many people in China are materially better off thanks to this policy, although some have gained much greater wealth than others.

The second priority was to ensure that there would never be a return to the chaos and terror of the Cultural Revolution that Deng and his colleagues had experienced. The result has been the maintenance of an authoritarian system, albeit with political reforms, and the continuing monopoly of political power and control by the Communist Party. The two priorities are conjoined in a new world view of the global marketplace and domestic political stability. This suits the trading and investor nations that are competing for a share of potentially the largest market in the world, exploiting cheap labor, unencumbered by regulations, in a relatively secure political environment.

China is reaching out, not only for export and investment opportunities, but also for international recognition and power. On the diplomatic front it has adopted a constructive stance on the Iraq conflict within the United Nations, and it is backing India's claim for a permanent seat on the UN Security Council. Within its own region it has taken the leading role in the multilateral attempts to seek a resolution to North Korea's nuclear posturing, and no longer remains aloof from negotiations on regional economic integration and security.

In 2004 the total value of China's trade was 30% higher than in 2003.

Militarily, mainland China and Taiwan are at a standoff, and the presence of the USA in the region looms large. But China is not about to challenge the USA's dominance either globally or regionally. At the end of 2004 it was, however, in the peculiar position of being the USA's indirect financial backer. The dollars China had earned as a result of its trade surplus were invested in US government securities and in institutions from which the USA, with its huge budget deficit, was borrowing billions.

China still faces obstacles. The 2008 Olympics is focusing attention on China's human rights record, and the so-called "war on terror" brings China's separatist movements into the spotlight. The suppression of Uygur groups for their mainly non-violent protests against the privileges of the Han Chinese, and the Party-state's control of Islamic education and practices, has raised concerns. The world will be watching to see how China's authoritarian system responds to the dilemma of its human rights record while reacting to possible terrorist threats.

WHEN AN ARROW IS ON A STRING IT MUST GO

In 1990, China was the 15th largest trading nation in the world. By 2003 it was the fourth, and snapping at the heels of Japan. If the Special Administrative Region of Hong Kong is taken into account, China moves up into second place. Despite importing billions of dollars of raw materials for use by its hungry manufacturing and industrial sectors, China is managing to hang on to an overall trade surplus.

China's economy grew at the extraordinary rate of over 8 percent per year between 2002 and 2004, and trade is playing a vital role in the country's economic liberalization and modernization. Between 2002 and 2003 China's total trade increased by $230 billion.

Entry into the World Trade Organization at the end of 2001 is responsible for an estimated $170 billion a year of this additional trade. However, China has been slow to lift import restrictions, while successfully exporting its own cheaply manufactured products worldwide. The bulk of China's trade, and the fastest-growing partnership, is with its Asian neighbors, but European traders are having some success in penetrating China's markets – more so than their American counterparts.

Countries with trade deficit

USA $2,030

Countries with trade surplus

In 2003 the USA had a trade deficit of $581 billion. China had a surplus of $26 billion.

$1,350 Germany

TOP 15 WORLD TRADERS
Total value of merchandise exports and imports
2003
US$ billion

Source: WTO

$855 Japan
$851 China ★

France $773

$664 Hong Kong ★

UK $692

$579 Italy
$555 Netherlands
$518 Canada
$489 Belgium

$373 South Korea

Mexico $344

$335 Singapore

$210 Russia

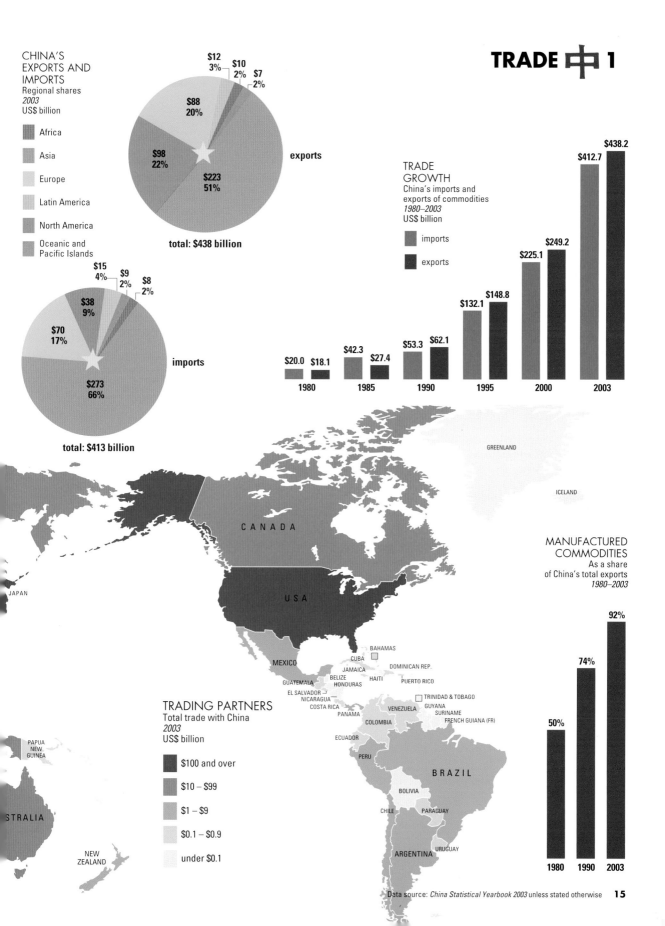

CHINA'S EXPORTS AND IMPORTS
Regional shares
2003
US$ billion

- Africa
- Asia
- Europe
- Latin America
- North America
- Oceanic and Pacific Islands

exports

$12 3%
$10 2%
$7 2%
$88 20%
$98 22%
$223 51%

total: $438 billion

imports

$15 4%
$9 2%
$8 2%
$38 9%
$70 17%
$273 66%

total: $413 billion

TRADE GROWTH
China's imports and exports of commodities
1980–2003
US$ billion

- imports
- exports

1980	1985	1990	1995	2000	2003
$20.0 $18.1	$42.3 $27.4	$53.3 $62.1	$132.1 $148.8	$225.1 $249.2	$412.7 $438.2

MANUFACTURED COMMODITIES
As a share of China's total exports
1980–2003

1980	1990	2003
50%	74%	92%

GREENLAND

ICELAND

JAPAN

CANADA

USA

BAHAMAS
CUBA
MEXICO
JAMAICA
DOMINICAN REP.
BELIZE
GUATEMALA HONDURAS HAITI PUERTO RICO
EL SALVADOR
NICARAGUA TRINIDAD & TOBAGO
COSTA RICA VENEZUELA GUYANA
PANAMA SURINAME
COLOMBIA FRENCH GUIANA (FR)
ECUADOR

PERU

BRAZIL

BOLIVIA

CHILE PARAGUAY

ARGENTINA URUGUAY

PAPUA NEW GUINEA

STRALIA

NEW ZEALAND

TRADING PARTNERS
Total trade with China
2003
US$ billion

- $100 and over
- $10 – $99
- $1 – $9
- $0.1 – $0.9
- under $0.1

Data source: *China Statistical Yearbook 2003* unless stated otherwise **15**

JUMPING INTO THE SEA

Foreign direct investment (FDI) in China is booming. The value contracted more than doubled between 1998 and 2003. Overseas companies are tempted by the liberalization of China's foreign-investment policies, its economic growth rate of over 8 percent a year, and the sheer size of its market: 1.3 billion people. And there is plenty of scope for further investment. Investors so far have focused mainly on the eastern coastal provinces, but there are signs that they are now looking hard at China's hinterland.

China's World Trade Organization membership is enabling foreign firms to enter into partnership with Chinese companies. Manufacturing is the most popular sector, with several of the world's largest car manufacturers opening factories, but real estate and the retail trade are also high on the list for investors.

Not all new entrepreneurial ventures run smoothly, however. Factories have been hit by problems with power supply, and everybody doing business in China has to be prepared to operate within China's own business code, which can attach greater importance to personal contacts than to legal documents. There is also the issue of copyright piracy, which is rampant in China and is making foreign investors wary. US, Japanese and European firms are estimated to lose around $50 billion annually through the illegal copying of pharmaceuticals, manufactured goods, books, films and music.

RECIPIENTS OF
FOREIGN DIRECT
INVESTMENT (FDI)
IN EAST ASIA AND
SOUTHEAST ASIA
2002
US$ billion

Source: *World Development Indicators 2004*

The Closer Economic Partnership Arrangement (CEPA), which came into effect on January 1, 2004, makes it easier for Hong Kong companies, and Hong Kong-based subsidiaries, to invest in mainland China.

$49.3 China ★

$12.7 Hong Kong ★

$9.1 Japan

$6.1 Singapore

$3.0 India
$2.0 South Korea
$1.4 Vietnam
$1.1 Philippines
$0.9 Thailand

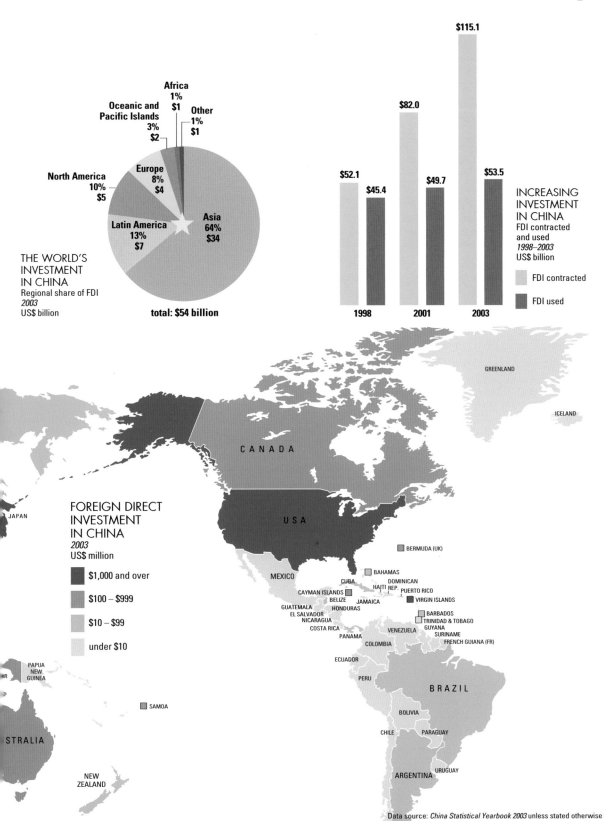

THE WORLD'S INVESTMENT IN CHINA
Regional share of FDI
2003
US$ billion

Africa
1%
$1

Oceanic and Pacific Islands
3%
$2

Other
1%
$1

North America
10%
$5

Europe
8%
$4

Latin America
13%
$7

Asia
64%
$34

total: $54 billion

$115.1

$82.0

$52.1

$45.4

$49.7

$53.5

INCREASING INVESTMENT IN CHINA
FDI contracted and used
1998–2003
US$ billion

FDI contracted

FDI used

1998   2001   2003

FOREIGN DIRECT INVESTMENT IN CHINA
2003
US$ million

$1,000 and over

$100 – $999

$10 – $99

under $10

GREENLAND

ICELAND

JAPAN

CANADA

USA

BERMUDA (UK)

BAHAMAS

MEXICO

CUBA

DOMINICAN
HAITI REP. PUERTO RICO

CAYMAN ISLANDS

BELIZE JAMAICA VIRGIN ISLANDS

GUATEMALA
EL SALVADOR HONDURAS
NICARAGUA

BARBADOS
TRINIDAD & TOBAGO

COSTA RICA GUYANA

PANAMA VENEZUELA SURINAME

COLOMBIA FRENCH GUIANA (FR)

ECUADOR

PERU

BRAZIL

BOLIVIA

CHILE PARAGUAY

ARGENTINA URUGUAY

PAPUA
NEW
GUINEA

SAMOA

STRALIA

NEW
ZEALAND

Data source: *China Statistical Yearbook 2003* unless stated otherwise   **17**

THE MASTER DOES NOT FIGHT; WERE HE TO DO SO HE WOULD WIN

China's regular armed forces number 2.25 million, comprise 11 percent of the world's total, and are nearly equal in size to the combined regular armies of Africa, and Central and South America. In addition, there is a sizeable reserve force of over half a million.

An army of this size is impressive on paper, and reasonably cheap to run, but in modern warfare it is high-tech weaponry that counts, and China's military expenditure is a long way behind that of the USA and the rest of NATO. China does have a nuclear capacity, however. While minuscule in comparison to the stockpiles of the USA and Russia, it is significant in terms of China's military power within Asia and Southeast Asia.

China has become increasingly keen to play a part on the world stage. It participated in its first United Nations peacekeeping operation in April 1992, and by the end of 2003 had dispatched more than 2,000 personnel, including engineers and medical staff, to missions around the world.

MILITARY EXPENDITURE
Comparative expenditure on defense
2003 selected countries
US$ billion

Source: *China Statistical Yearbook 2003*

$405 USA

$211 NATO Europe

$65 Russia

$56 China ★

$46 France
$43 Japan, UK
$35 Germany
$28 Italy
$16 India
$15 South Korea
$12 Australia

CHINA'S PEACEKEEPING CONTRIBUTION
2004
Countries with UN peacekeeping operations to which China has contributed personnel

current UN peacekeeping operations

past UN peacekeeping operations

Source: www.un.org

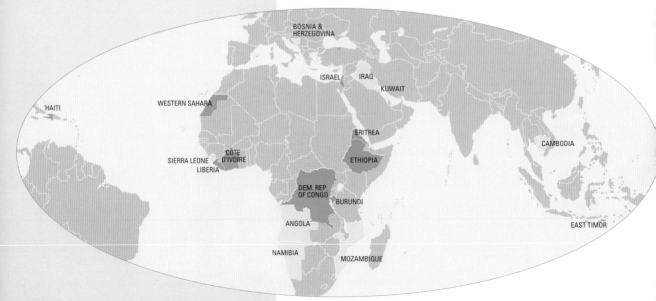

BOSNIA & HERZEGOVINA

ISRAEL IRAQ

KUWAIT

HAITI

WESTERN SAHARA

ERITREA

CAMBODIA

SIERRA LEONE CÔTE D'IVOIRE

ETHIOPIA

LIBERIA

DEM. REP. OF CONGO

BURUNDI

ANGOLA

EAST TIMOR

NAMIBIA

MOZAMBIQUE

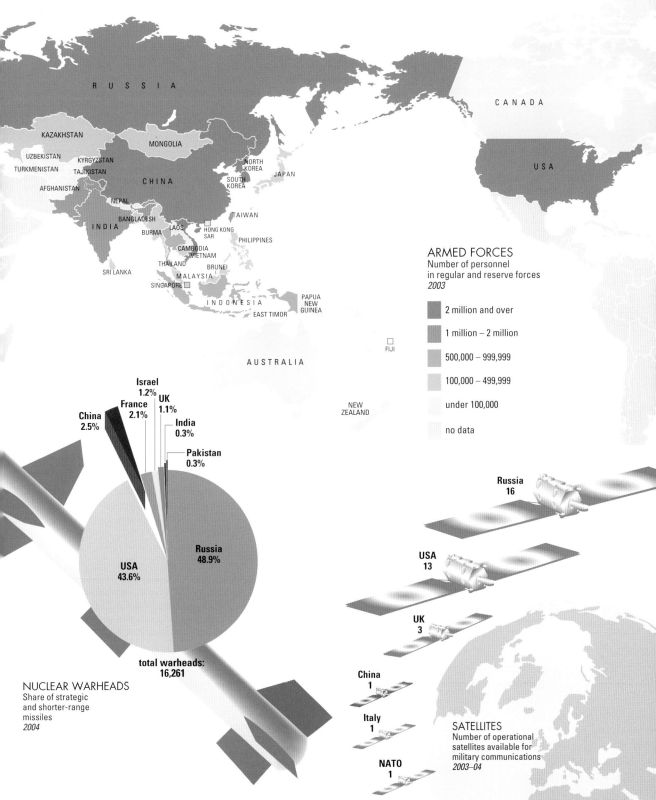

ARMED FORCES
Number of personnel
in regular and reserve forces
2003

- 2 million and over
- 1 million – 2 million
- 500,000 – 999,999
- 100,000 – 499,999
- under 100,000
- no data

Israel
1.2%
France
2.1%
UK
1.1%
China
2.5%
India
0.3%
Pakistan
0.3%
Russia
48.9%
USA
43.6%

total warheads:
16,261

NUCLEAR WARHEADS
Share of strategic
and shorter-range
missiles
2004

Russia
16
USA
13
UK
3
China
1
Italy
1
NATO
1

SATELLITES
Number of operational
satellites available for
military communications
2003–04

Data source: IISS, *The Military Balance 2004–05* unless stated otherwise

THE FIRST UNDER HEAVEN

China's relations with the rest of world were thrown into disarray by the collapse of European communism and the break-up of the Soviet Union. In the new world order of the 1990s, China had to make its own way, somewhat hampered by the negative fall-out from its repression of democracy supporters in Tiananmen Square in 1989. The path has not always been smooth, but it has had a clear direction: integration into the global economy through trade and investment, and the political clout that being a major economic player brings.

Global aspirations are, however, tempered by regional interests. China has occupied a permanent seat on the United Nations Security Council since 1971, when it took over the seat occupied by Taiwan. Unlike the other permanent members – France, Russia, the UK and USA – it has used its power to veto Security Council resolutions sparingly, but moves to award Japan a permanent seat may well prompt China into exercising that power.

China sees Taiwan as a renegade province, while a majority of Taiwan's people regard their island as a sovereign state. This leads to a fraught relationship, but potential flashpoints have so far been managed, and conflict avoided. While international attention focuses on the military threat China poses to Taiwan, the smaller country is gaining in international stature from its robust democratic politics and economic progress, which includes an investment of more than $50 billion in China's industry.

China–European Union (EU)

- **1998** China and EU countries propose closer ties.
- **2000** EU asks China to take action against 400 Chinese companies involved in theft of intellectual property rights.
- **2001–02** EU puts pressure on China to improve its human rights record.
- **2001–04** China–EU trade relations are further facilitated.
- EU takes legal action against China's "dumping" of cheap products on EU markets, and its breach of intellectual property rights.

CHINA'S TERRITORIAL CLAIMS
2004

countries laying territorial claim to Spratly Islands or to an area that includes them

Source: www.eia.doe.gov/emeu/cabs/schina.html

China claims an extensive area of the South China Sea, in contravention of international law, as does Taiwan. This includes more than 200 tiny islands, rocks and reefs. In 1974 Chinese troops seized the Paracel Islands from Vietnam. The Spratly Islands are even more hotly contested. Troops from China, Malaysia, Taiwan and the Philippines have occupied some of the islands, and since 1988 there have been a number of military clashes.

Legal ownership of the islands would confer the right to exploit the extensive natural resources under the surrounding sea bed. China signed a contract with the US firm Crestone in 1992 and sank its first exploratory drill west of the Spratly Islands in that year. It continues to extract oil from the area, despite protests from Vietnam. In 2002 China signed, and subsequently ratified, the Treaty of Amity, a disputes procedure established by the Association of Southeast Asian Nations (ASEAN). This may prove a positive step towards reconciling differences in the region, or a cynical ploy by China, who, at the same time as making diplomatic moves, has established a strong military presence on the two island groups.

China–Russia and Central Asia

- **1991–92** China establishes diplomatic relations with newly independent ex-Soviet states.
- **1996 onwards** Russia and China talk of "strategic partnership".
- **2001** China signs up to revitalized Shanghai Cooperation Organization, along with Russia, Kazakhstan, Kyrgyzstan, Tajikistan and Uzbekistan.
- **2003** President Hu visits Russia. Makes agreement for pipeline to transport Siberian oil to China.
- **2004** China settles border dispute with Russia.

CHINA'S CHANGING INTERNATIONAL RELATIONS
1989–2004

- ● action detrimental to China
- ○ action beneficial to China

China–Asia

- ● **1995** China's occupation of Mischief Reef in South China Sea disputed by Philippines.
- ● **1996** China undertakes military exercises in Taiwan Straits, including launching missiles during run-up to Taiwan's elections.
- ● **1998** President Jiang Zemin visits Tokyo but fails to win apology from Japan for WWII.
- ○ **2000** Taiwanese politicians make unofficial visit to Beijing.
- ○ **2000** North Korean leader, Kim Jong-il, visits Beijing prior to summit with South Korean leader.
- ○ **2000** China and Vietnam settle dispute over joint border and Gulf of Tonkin.
- ● **2001** (August) Japanese PM Koizumi makes first visit to shrine to honor Japan's war dead.
- ○ **2002** (Jan) First Chinese state visit to India for over a decade.
- ○ **2002** (Nov) Premier Zhu Rongji signed commitment with Association of Southeast Asian Nations (ASEAN) to form free-trade area by 2010, and to prevent clashes in disputed areas in South China Sea.
- ○ **2003** (April) China sponsors trilaterial talks with North Korea and USA.
- ● **2003** (Feb) China shuts down oil supplies to North Korea.
- ● **2004** (January) Koizumi again visits shrine to honor Japan's war dead.

China–USA

- ● **1989** Following suppression of Tiananmen Square protest, arms embargo imposed by USA and EU.
- ○ **1989–1992** US President George Bush boosts trade with China.
- ○ **1990–91** China sides with US-led coalition in run-up to first Gulf War, but abstains from authorizing use of force.
- ● **1993** Clinton links trade with human rights record, to China's disadvantage.
- ○ **1994** Clinton revives trading links with China, regardless of human rights record.
- ● **1995** China protests when Taiwan President Lee Tung-Hui visits USA.
- ○ **1995** Clinton tells President Jiang Zemin of the 'Three Nos': No recognition of Taiwan independence; No support for two Chinas; No support for Taiwan's entry into international organizations.
- ● **1996** China threatens Taiwan with missile tests. USA sends aircraft carriers to prevent hostilities.
- ○ **1997** Jiang Zemin visits USA.
- ○ **1998** Clinton visits China. He publicly repeats the 'Three Nos'.
- ● **1999** (June) Accidental bombing by US of China's Belgrade Embassy during Kosovan crisis.
- ○ **1999** (Nov) US and China announce pacts on terms for China's entry to World Trade Organization.
- ○ **2000** (Oct) Clinton signs law giving US normal trade relations with China.
- ● **2001** Collision between Chinese jetfighter and US spyplane.
- ○ **2001** China supportive of US actions in Afghanistan.
- ○ **2002** President George W Bush visits Beijing; General Secretary Hu Jintao visits USA. USA adds East Turkestan Islamic Movement to its list of terrorist organizations. China votes for UN ultimatum to Iraq.
- ○ **2004** US Secretary of State Colin Powell visits Beijing and strengthens US commitment to One-China Policy.

China–Central and South America

- ○ **2004** President Hu Jintao visits Argentina, Brazil, Chile and Cuba. Trade is expected to increase 150% by 2010.

Part Two

CHINA'S PEOPLE

CHINA IS A HUGELY POPULOUS NATION and for many years that single fact has been the undisputed characteristic of both its potential and its challenge. Yet, any understanding of China must also take into account particular sectors of the population, discrete interest groups, those that are under-represented in the process of modernization, and those that dominate the policy agenda.

In Chinese research, this is usually handled by insisting on a comparative, or at least delineated, approach to the "rural" and the "urban". This separation is echoed in the data collected in the Chinese statistical yearbooks and in the registration system for citizens, and is derived from longstanding divisions between an urban proletariat (and intelligentsia) and the peasants. Recently, there have been efforts to re-define the ways in which China is understood, partly by scholarly efforts to reassess provinces as important units of experience, economy and culture, and partly by the emphasis placed by the Chinese government on macro-regions for planning.

"It is easier to find Chinese-ness rooted in history than in the shared qualities among people known as Chinese."
Wang Guangwu

The eastern seaboard, the central area and the west have become useful shorthand for stages of development. The east is seen as predominantly developed, densely populated, urban, and Han, and the west is understood as underdeveloped, populated by minority ethnicities as well as Han migrants, with semi-rural societies continuing to dominate. One explanation for this is geographical. The seaboard has always had access to international trade, and so has a population used to change. The changes that have occurred in the land-locked west have been more political in nature, based on the area's proximity – both ethnic and geographic – to new states formed after the demise of the Soviet Union.

Unsurprisingly, the north-central region sits between these two extremes, with some pressing developmental issues, and a mix of Han and minority social structures moving towards modernization in provincially specific ways.

The Han Chinese population group is in the overwhelming majority, and the very term "Chinese" presupposes many Han ways, beliefs and historical references, but also many that are originally from other ethnic groups – or which have been entirely made up or re-invented. Indeed, one could argue that "Chinese-ness" is an artificial and constructed idea, always open to negotiation, and dependent on language, cultural practices and place of residence. Arguably, shared qualities are as likely to be found in the border cities of Tibet and Xinjiang, amongst people who are ethnically diverse but geographically proximate, as they are amongst Han Chinese across the whole of China and beyond.

Chinese commentators are also concerned with the gender divide. In general, the one-child policy – an effective although crude tool of population management – has not had an adverse impact on the relative numbers of urban women in the higher socio-economic stratas. Amongst rural and less affluent groups, however, the need for male children has caused girl babies to suffer neglect, abandonment and elective abortion.

This is not an aim of the central government, and indeed causes great concern. Nevertheless, government policies lead to unexpected and unhappy results. The re-privatization of responsibility for agricultural work, within a social system in which girls still move away on marriage, requires male input on the land and prejudices parents in favor of a son. Even in urban environments some unfortunate women find that the re-introduction of "liberal" values enables men to desert them to find another – usually younger – woman to provide a male offspring when the first try was unsuccessful.

YOU CANNOT WRAP A FIRE IN PAPER

Nearly 1.3 billion people, over one fifth of the world's population, live in China. The sheer numbers involved affect all aspects of life. The population continues to increase – even though the rates of growth have slowed. As the population clock suggests, these numbers challenge available solutions.

China's population is unevenly distributed across its provinces. Urban areas are becoming more overcrowded as the rural population leaves the land to work in the towns and cities, especially those in the eastern region.

In 2000, China conducted the world's largest census. Despite the difficulties in taking an accurate account of such large numbers, and ensuring the cooperation of local officials, it was pronounced a great success.

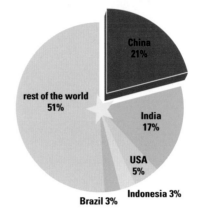

WORLD POPULATION
Share of total population
2004

World population: 6.2 billion

Source: *World Development Indicators 2004*

China 21%
rest of the world 51%
India 17%
USA 5%
Indonesia 3%
Brazil 3%

In China there are four times as many people as in the USA.

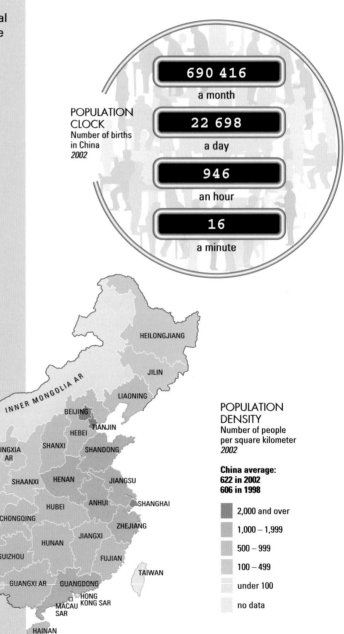

POPULATION CLOCK
Number of births in China
2002

690 416	a month
22 698	a day
946	an hour
16	a minute

POPULATION DENSITY
Number of people per square kilometer
2002

China average:
622 in 2002
606 in 1998

- 2,000 and over
- 1,000 – 1,999
- 500 – 999
- 100 – 499
- under 100
- no data

HEILONGJIANG
JILIN
LIAONING
XINJIANG AR
INNER MONGOLIA AR
GANSU
BEIJING
TIANJIN
HEBEI
QINGHAI
NINGXIA AR
SHANXI
SHANDONG
SHAANXI
HENAN
JIANGSU
TIBET AR
ANHUI
SHANGHAI
SICHUAN
HUBEI
CHONGQING
ZHEJIANG
JIANGXI
HUNAN
GUIZHOU
FUJIAN
TAIWAN
YUNNAN
GUANGXI AR
GUANGDONG
HONG KONG SAR
MACAU SAR
HAINAN

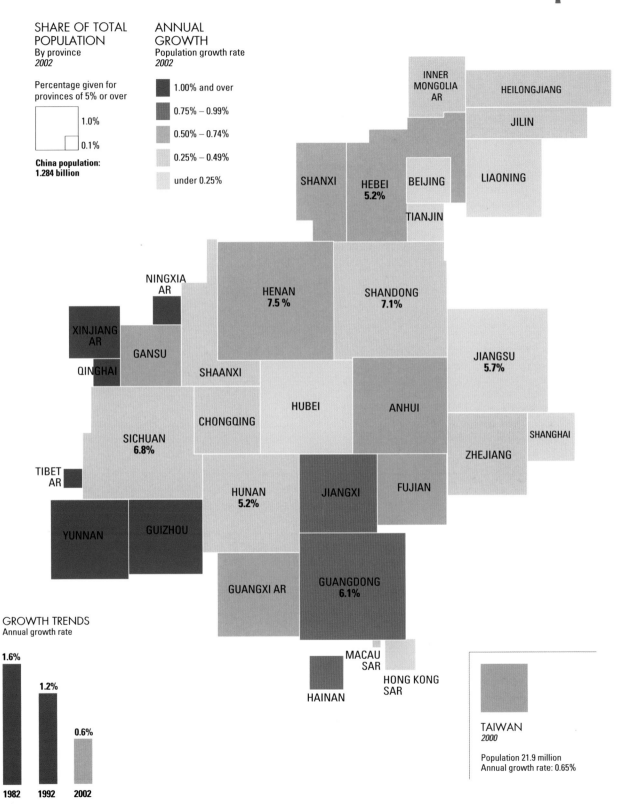

SHARE OF TOTAL POPULATION
By province
2002

Percentage given for provinces of 5% or over

1.0%
0.1%

China population: 1.284 billion

ANNUAL GROWTH
Population growth rate
2002

- 1.00% and over
- 0.75% – 0.99%
- 0.50% – 0.74%
- 0.25% – 0.49%
- under 0.25%

INNER MONGOLIA AR

HEILONGJIANG

JILIN

SHANXI

HEBEI
5.2%

BEIJING

LIAONING

TIANJIN

NINGXIA AR

HENAN
7.5 %

SHANDONG
7.1%

XINJIANG AR

GANSU

JIANGSU
5.7%

QINGHAI

SHAANXI

HUBEI

ANHUI

CHONGQING

SHANGHAI

SICHUAN
6.8%

ZHEJIANG

TIBET AR

HUNAN
5.2%

JIANGXI

FUJIAN

YUNNAN

GUIZHOU

GUANGXI AR

GUANGDONG
6.1%

MACAU SAR

HONG KONG SAR

HAINAN

GROWTH TRENDS
Annual growth rate

1.6%
1.2%
0.6%

1982 1992 2002

TAIWAN
2000

Population 21.9 million
Annual growth rate: 0.65%

MONEY CAN MOVE EVEN THE GODS

By 2020 China is expected to be a predominantly urban society, with maybe 60 percent of its people living in towns and cities.

From 1997 to 2002 China's rural population declined by seven percent. People have been displaced from rural areas by the de-collectivization of agriculture, and many have been lured to the cities in search of a better living. Rural migrants (*mingong*) form the bulk of the labor force in the massive construction projects taking place in cities such as Shanghai. They work for long hours and low wages but, by living in cramped and basic conditions, save money to send home to their villages. In this way, the effects of the east-coast economic boom are slowly trickling through to the rural areas.

Other less fortunate rural migrants are unable to find work in cities. The system of registration that determines someone's official place of residence (*hukou*) has made it difficult for these people to claim housing and welfare benefits. However, there are signs that controls are being relaxed, which will make migration easier.

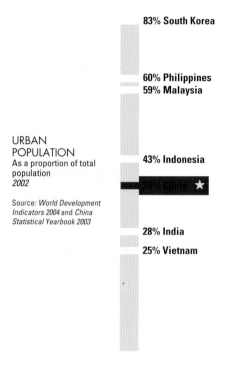

URBAN POPULATION
As a proportion of total population
2002

Source: *World Development Indicators 2004* and *China Statistical Yearbook 2003*

83% South Korea
60% Philippines
59% Malaysia
43% Indonesia
39% China ★
28% India
25% Vietnam

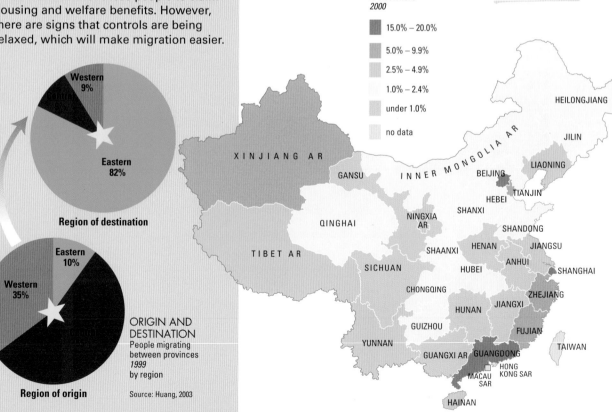

Western 9%

Eastern 82%

Region of destination

Eastern 10%

Western 35%

Region of origin

ORIGIN AND DESTINATION
People migrating between provinces
1999
by region

Source: Huang, 2003

MIGRATION
Percentage of population resident in province whose official residence is elsewhere
2000

- 15.0% – 20.0%
- 5.0% – 9.9%
- 2.5% – 4.9%
- 1.0% – 2.4%
- under 1.0%
- no data

144 million people have left rural areas in recent years to seek work in cities.

XINJIANG AR
GANSU
INNER MONGOLIA AR
BEIJING
HEILONGJIANG
JILIN
LIAONING
TIANJIN
HEBEI
SHANXI
QINGHAI
NINGXIA AR
SHANDONG
SHAANXI
HENAN
JIANGSU
TIBET AR
SICHUAN
HUBEI
ANHUI
SHANGHAI
CHONGQING
ZHEJIANG
HUNAN
JIANGXI
GUIZHOU
FUJIAN
TAIWAN
YUNNAN
GUANGXI AR
GUANGDONG
HONG KONG SAR
MACAU SAR
HAINAN

China has 166 cities with populations over 1 million; the USA has just nine.

URBAN POPULATION
As a percentage of total population of province
2000

- 75% and over
- 50% – 74%
- 25% – 50%
- under 25%
- no data

Source: Lavely, 2001

URBAN GROWTH
Percentage increase in urban population
1990–2000

- ⬆ 20% and over
- ⬆ 10% – 19%

Source: Lavely, 2001

Map labels

HEILONGJIANG

JILIN

LIAONING

XINJIANG AR

INNER MONGOLIA AR

GANSU

BEIJING

TIANJIN

HEBEI

NINGXIA AR

SHANXI

SHANDONG

QINGHAI

SHAANXI

HENAN

JIANGSU

TIBET AR

ANHUI

SHANGHAI

SICHUAN

CHONGQING

HUBEI

ZHEJIANG

JIANGXI

HUNAN

GUIZHOU

YUNNAN

GUANGXI AR

GUANGDONG

FUJIAN

TAIWAN

HONG KONG SAR

MACAU SAR

HAINAN

CHANGING URBAN–RURAL BALANCE
Percentage of population living under urban and rural administrations
1978–2002

- urban
- rural

Year	urban	rural
1978	18%	82%
1989	26%	74%
1997	32%	68%
2002	39%	61%

CHANGING CITY SIZE
Number of cities with non-agricultural population over 1 million
1980–2001

Source: *China Population Statistics Yearbook 2002*

Year	Number
1980	15
1990	31
2001	41

REARING A TIGER
IS TO INVITE FUTURE TROUBLE

There are more boys than girls in China. This is particularly true in the rural areas, where boys are seen as more productive in agricultural work, and more valuable to aging parents. The consequences of small- or one-child family policies has been severe for girl children in the countryside. There, stories of abandonment and neglect are common, and tales of abductions – of adolescent girls and young women – suggest a widening gender gap amongst under 30-year-olds.

Implementation of the law is left to provincial governments, who may vary it according to local conditions. Concerned about a lack of people to care for its increasingly elderly population, Shanghai has relaxed its regulations to allow some couples to have two children, and now offers incentives to daughter-only families.

Some rural women have been subject to forced sterilization in certain provinces – the same areas that now have the worst gender ratio.

Female sterilization represents a 68% share of contraception in Gansu, compared with only 3% in Shanghai.

Use of the IUD increased by 1% during 2000–01, at the expense of sterilization.

GENDER INEQUALITY
Gender-related Development Index (GDI) score
2004

The GDI is a combined measurement of life expectancy, literacy and earnings for women, compared with those for men. The higher the score, the more equal the society.

most equal
.945 Australia
.936 USA
.934 UK
.929 France

.794 Russia
.768 Brazil
.741 China ★
.685 Indonesia

.572 India

.458 Nigeria

least equal

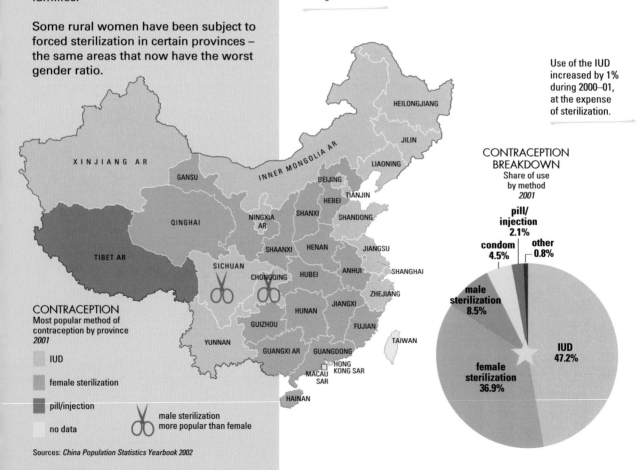

CONTRACEPTION
Most popular method of contraception by province
2001

- IUD
- female sterilization
- pill/injection
- no data

✂ male sterilization more popular than female

CONTRACEPTION BREAKDOWN
Share of use by method
2001

pill/injection 2.1%
condom 4.5%
other 0.8%
male sterilization 8.5%
IUD 47.2%
female sterilization 36.9%

Sources: *China Population Statistics Yearbook 2002*

The sex ratio imbalance is due to the under-reporting of female births, especially in the countryside, aborting more female than male foetuses, and female infanticide.

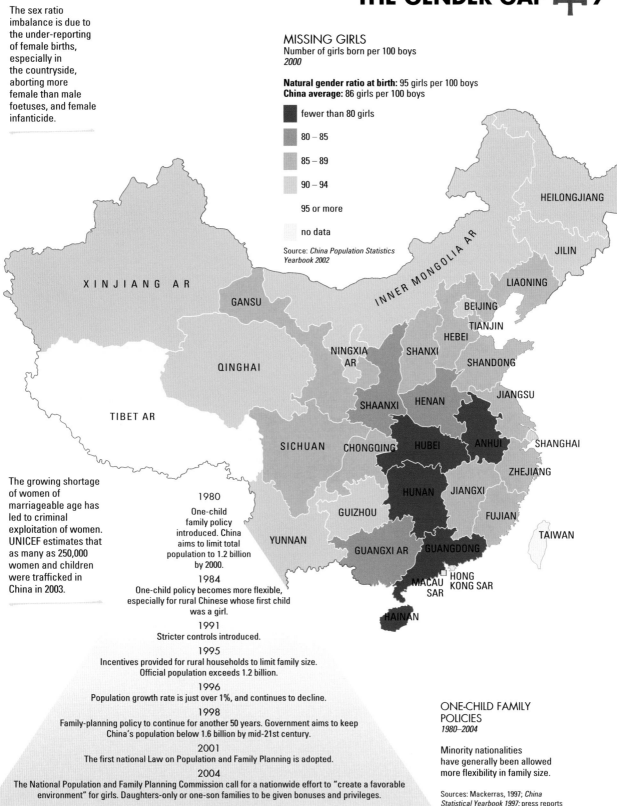

MISSING GIRLS
Number of girls born per 100 boys
2000

Natural gender ratio at birth: 95 girls per 100 boys
China average: 86 girls per 100 boys

- fewer than 80 girls
- 80 – 85
- 85 – 89
- 90 – 94
- 95 or more
- no data

Source: *China Population Statistics Yearbook 2002*

The growing shortage of women of marriageable age has led to criminal exploitation of women. UNICEF estimates that as many as 250,000 women and children were trafficked in China in 2003.

1980
One-child family policy introduced. China aims to limit total population to 1.2 billion by 2000.

1984
One-child policy becomes more flexible, especially for rural Chinese whose first child was a girl.

1991
Stricter controls introduced.

1995
Incentives provided for rural households to limit family size. Official population exceeds 1.2 billion.

1996
Population growth rate is just over 1%, and continues to decline.

1998
Family-planning policy to continue for another 50 years. Government aims to keep China's population below 1.6 billion by mid-21st century.

2001
The first national Law on Population and Family Planning is adopted.

2004
The National Population and Family Planning Commission call for a nationwide effort to "create a favorable environment" for girls. Daughters-only or one-son families to be given bonuses and privileges.

ONE-CHILD FAMILY POLICIES
1980–2004

Minority nationalities have generally been allowed more flexibility in family size.

Sources: Mackerras, 1997; *China Statistical Yearbook 1997;* press reports

WHEN THE NEST IS OVERTURNED NO EGG STAYS UNBROKEN

When the Chinese Communist Party came to power in 1949, it changed one component of the characters for ethnic names: the symbol for "dog" was replaced by the symbol for "man". Despite this auspicious start, ethnic minorities have not always been comfortable within the territory of the People's Republic of China. The Tibetans and minority nationalities in Xinjiang actively work for separation from China. The Hakka are still waiting for minority nationality status.

Because of the strong presence of certain minority nationalities in Guangxi, Inner Mongolia, Ningxia, Tibet, and Xinjiang, these have been designated Autonomous Regions (ARs). This confers national minorities with some political and cultural rights, but, in practice, they enjoy little power. Several ARs are located along China's borders, and are significant in terms of national security. Others are rich in natural resources, and vital to the country's economy. Not surprisingly, therefore, all secessionist activities are banned.

During the late 1990s, the Dalai Lama's demands for full independence were modified to "genuine self-rule".

The first railway link between Lhasa in Tibet and Qinghai is due for completion in 2007.

There are nearly 100,000 Tibetans living in India and 20,000 in Nepal.

TOTAL POPULATIONS OF THE MAJOR MINORITY NATIONALITIES
2000
Populations of 1 million or more

Zhuang	16.2 m
Manchu	10.7 m
Hui	9.8 m
Miao	8.9 m
Uygur	8.4 m
Tujia	8.0 m
Yi	7.8 m
Mongolian	5.8 m
Tibetan	5.4 m
Bouyei	3.0 m
Dong	3.0 m
Yao	2.6 m
Korean	1.9 m
Bai	1.9 m
Hani	1.4 m
Kazak	1.3 m
Li	1.2 m
Dai	1.2 m

AUTONOMOUS REGIONS (ARs)
and when established

There are also 30 autonomous prefectures, 120 autonomous counties and 1,300 ethnic minority townships

Source: State Council Information Office, 2003

XINJIANG-UYGUR AR **1955**

INNER MONGOLIA AR **1947**

NINGXIA-HUI AR **1958**

TIBET AR **1965**

GUANGXI ZHUANG AR **1958**

XINJIANG

GANSU
367,000

QINGHAI
900,000

TIBET
2 m

Lhasa •

SICHUAN
1 m

YUNNAN
111,000

TIBETANS

Tibet Autonomous Region

other areas with Tibetan autonomous status

area claimed by Tibetan government in exile

total population of Tibetans by province *1995*

Sources: Barnett, 1994; *China Population Statistics Yearbook 1997*; Mackerras, 1997; press reports

MINORITY NATIONALITIES 中 8

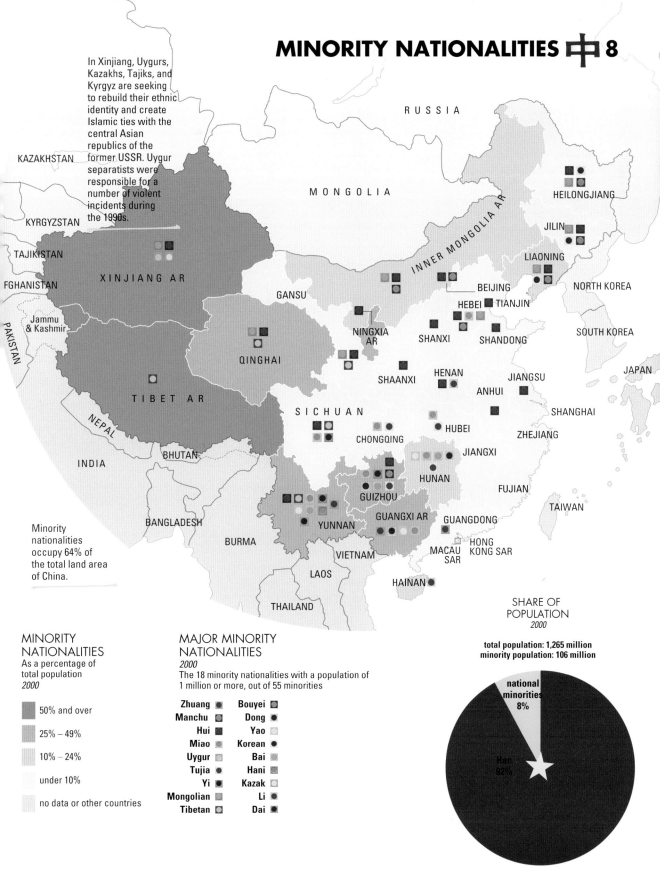

In Xinjiang, Uygurs, Kazakhs, Tajiks, and Kyrgyz are seeking to rebuild their ethnic identity and create Islamic ties with the central Asian republics of the former USSR. Uygur separatists were responsible for a number of violent incidents during the 1990s.

Minority nationalities occupy 64% of the total land area of China.

RUSSIA

MONGOLIA

KAZAKHSTAN

KYRGYZSTAN

TAJIKISTAN

AFGHANISTAN

PAKISTAN

Jammu & Kashmir

XINJIANG AR

GANSU

INNER MONGOLIA AR

HEILONGJIANG

JILIN

LIAONING

NORTH KOREA

SOUTH KOREA

JAPAN

BEIJING

HEBEI   TIANJIN

NINGXIA AR

SHANXI   SHANDONG

QINGHAI

SHAANXI

HENAN

ANHUI

JIANGSU

SHANGHAI

TIBET AR

NEPAL

SICHUAN

CHONGQING

HUBEI

ZHEJIANG

BHUTAN

INDIA

JIANGXI

HUNAN

FUJIAN

TAIWAN

BANGLADESH

GUIZHOU

YUNNAN

GUANGXI AR

GUANGDONG

MACAU SAR

HONG KONG SAR

BURMA

VIETNAM

LAOS

HAINAN

THAILAND

MINORITY NATIONALITIES
As a percentage of total population
2000

- 50% and over
- 25% – 49%
- 10% – 24%
- under 10%
- no data or other countries

MAJOR MINORITY NATIONALITIES
2000

The 18 minority nationalities with a population of 1 million or more, out of 55 minorities

Zhuang	Bouyei	
Manchu	Dong	
Hui	Yao	
Miao	Korean	
Uygur	Bai	
Tujia	Hani	
Yi	Kazak	
Mongolian	Li	
Tibetan	Dai	

SHARE OF POPULATION
2000

total population: 1,265 million
minority population: 106 million

national minorities 8%

Han 92%

Data source: *China Statistical Yearbook 2003* unless stated otherwise   **31**

Part Three

THE ECONOMY

THE CHINESE ECONOMY is central to global financial development, but it is neither predictable nor static. In the past 25 years, China has shifted from being seen as an opportunity for the expansion of corporate capitalism, to being lauded as the next knowledge economy. Its rapid economic growth has thrust it onto the world stage, causing it to become the pre-eminent regional power and a major player on the world's trading floors, as well as giving it influence over international exchange rates. Alongside this runaway success story, however, alarm bells are sounding over the dangers to China's social structure of such rapid and fundamental change. It is a fascinating conundrum.

China has become an entrepreneurial state, with individuals striving to meet heightened expectations through market-oriented schemes, and mindful of the need to amass individual savings in order to offset a declining welfare provision. Individual wealth is rising sharply, and with it the market for luxury private cars and domestic labor in brand-new homes. China is aiming to become a "connected" nation, with improved access to transport allowing the "opening up" of the Western Provinces and Autonomous Regions, and domestic tourism reaching unprecedented levels. Agriculture is also changing. Many more farmers are turning away from the staples of food security, grain and rice, and planting orchards, vegetable crops, and raising more animals and fish. They are responding to changing times: local demands for a varied diet, the appetites of fast-food chains, and the lucrative potential of export markets.

Accession to the World Trade Organization at the end of 2001 was supposed to integrate China's practices into the world economy, aligning the largest socialist nation on earth with the interests and paradigms of a capitalist corporate management. The Chinese Party-state sees the WTO as a "wrecking ball" that will sort out the weak from the strong with as few political ramifications as possible. The internationally devised WTO regulations can be blamed for any internal pain, whilst the state can applaud its own successes.

For those who benefit, the new economy is all good news. State intervention in the past produced a wasteful and uninspired range of goods and services. Accession to the WTO promises to ensure that the Chinese authorities are transparent in their purchasing and tendering from suppliers and improve their standards of accounting, banking and securities management.

In 2002 there were 64 million small investors in China's domestic stock market – only slightly fewer than the number of Chinese Communist Party members.

However, the rifts are deepening between those benefiting from the newly created wealth, symbolized by the gleaming buildings and bright lights of Shanghai's financial district, and the urban underclass and rural poor, who experience labor insecurity and unemployment. Alongside this social and economic inequality are the dangers of an agricultural sector driven by immediate profit, rather than long-term food security, and environmental degradation caused by excessive energy consumption and poor anti-pollution regimes.

These conditions are compounded by what seems increasingly to be the chronic state of transition in China's economic, social and political policies. No-one is yet sure where they will end up in the pecking order of new China, but all are sure that there is no benefit either in withdrawal from the race to wealth, or in submitting, as low-waged migrant workers, to other people's success.

BETTER THE HEAD OF A CHICKEN THAN THE TAIL OF AN OX

China is fond of designating heroes for emulation. During the Maoist period there was the trinity of worker (now employee), peasant (now farmer) and PLA fighter (now soldier). Today it is the entrepreneur, scion of the market-oriented economy. Following Deng Xiaoping's dictum, "Get Rich First", peasants moved out of grain production into specialist and market gardening, or set up town or village enterprises. The area of land used for vegetable growing increased by 80 percent between 1996 and 2003. Many people not only grew rich, but became village leaders, chairing the village enterprises management committees.

The reforms of the state-owned enterprises, the opening of China's economy to foreign investment and tourism and, more recently, the large-scale privatization of housing has created an entrepreneurial class. Most dramatic has been the rise of the property developer.

Modest entrepreneurs have also prospered. Market stalls sell copies of Mao Zedong badges and copies of the *Little Red Book*. The irony does not end there, however. At one time there were more than 40 theme restaurants in Beijing alone serving Mao Zedong's favorite dishes.

The China Entrepreneurial Confidence Index surveys 19,500 entrepreneurs in different types of enterprises, including state-owned, collectively owned, jointly owned and private enterprises.

FLUCTUATING ENTREPRENEURIAL CONFIDENCE
According to China National Census Bureau quarterly index
2001–2004

Source: www.stats.gov.cn
People's Daily Online

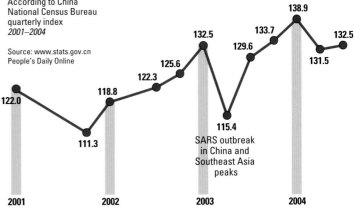

122.0
111.3
118.8
122.3
125.6
132.5
115.4
129.6
133.7
138.9
131.5
132.5

SARS outbreak in China and Southeast Asia peaks

2001   2002   2003   2004

SAVINGS DEPOSITS
Amount saved
1990–2002
billion yuan

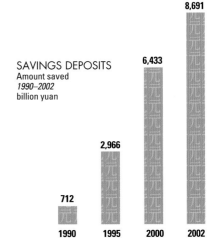

712
2,966
6,433
8,691

1990   1995   2000   2002

PROFILE OF FEMALE ENTREPRENEURS
2004

Source: People's Daily Online

20% of Chinese entrepreneurs are women. Over 95% of medium and small enterprises led by women were making a profit in 2004.

80% aged 30 – 50 years

56% have achieved associate college degree or above

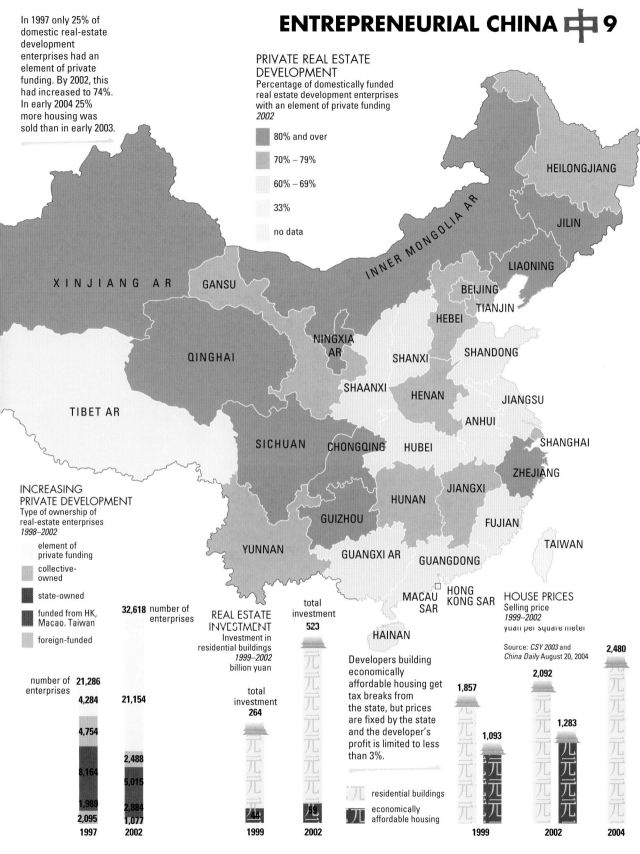

In 1997 only 25% of domestic real-estate development enterprises had an element of private funding. By 2002, this had increased to 74%. In early 2004 25% more housing was sold than in early 2003.

PRIVATE REAL ESTATE DEVELOPMENT
Percentage of domestically funded real estate development enterprises with an element of private funding
2002

- 80% and over
- 70% – 79%
- 60% – 69%
- 33%
- no data

HEILONGJIANG

JILIN

LIAONING

INNER MONGOLIA AR

XINJIANG AR

GANSU

BEIJING

TIANJIN

HEBEI

SHANDONG

NINGXIA AR

SHANXI

QINGHAI

SHAANXI

HENAN

JIANGSU

TIBET AR

ANHUI

SHANGHAI

SICHUAN

CHONGQING

HUBEI

ZHEJIANG

JIANGXI

HUNAN

GUIZHOU

FUJIAN

TAIWAN

YUNNAN

GUANGXI AR

GUANGDONG

MACAU SAR

HONG KONG SAR

HAINAN

INCREASING PRIVATE DEVELOPMENT
Type of ownership of real-estate enterprises
1998–2002

- element of private funding
- collective-owned
- state-owned
- funded from HK, Macao, Taiwan
- foreign-funded

number of enterprises

	1997	2002
	21,286	32,618
	4,284	21,154
	4,754	
		2,488
	8,164	5,015
	1,989	2,884
	2,095	1,077

REAL ESTATE INVESTMENT
Investment in residential buildings
1999–2002
billion yuan

total investment
264 (1999)

total investment
523 (2002)

- residential buildings
- economically affordable housing

1999: 44
2002: 59

Developers building economically affordable housing get tax breaks from the state, but prices are fixed by the state and the developer's profit is limited to less than 3%.

HOUSE PRICES
Selling price
1999–2002
yuan per square meter

Source: *CSY 2003* and *China Daily* August 20, 2004

1999: 1,857 / 1,093
2002: 2,092 / 1,283
2004: 2,480

LET SOME PEOPLE GET RICH FASTER THAN OTHERS

Despite the overall improvement in living standards that began with the economic reforms of 1978, serious disparities still exist between the urban and rural populations.

There is also a deep-seated inequality in terms of natural resources and financial wealth between the Central and Eastern regions, which are well-placed for economic development, and the mainly rural Western hinterlands. The party-state has been addressing this imbalance and is aiming to increase the income of rural populations by cutting taxes, cracking down on corrupt local officials and higher grain prices, and by improving farming methods.

In 2004 the growth rate for rural incomes was the highest for eight years.

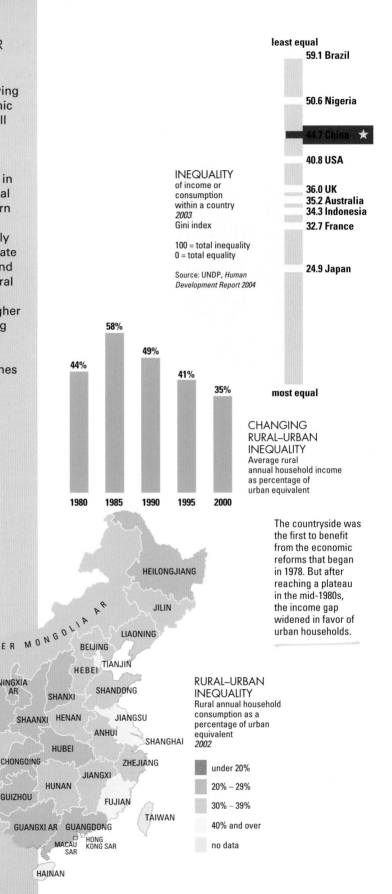

INEQUALITY
of income or consumption within a country
2003
Gini index

100 = total inequality
0 = total equality

Source: UNDP, *Human Development Report 2004*

least equal

59.1 Brazil

50.6 Nigeria

44.7 China ★

40.8 USA

36.0 UK
35.2 Australia
34.3 Indonesia
32.7 France

24.9 Japan

most equal

CHANGING RURAL–URBAN INEQUALITY
Average rural annual household income as percentage of urban equivalent

- 1980 — 44%
- 1985 — 58%
- 1990 — 49%
- 1995 — 41%
- 2000 — 35%

The countryside was the first to benefit from the economic reforms that began in 1978. But after reaching a plateau in the mid-1980s, the income gap widened in favor of urban households.

RURAL–URBAN INEQUALITY
Rural annual household consumption as a percentage of urban equivalent
2002

- under 20%
- 20% – 29%
- 30% – 39%
- 40% and over
- no data

In Tibet rural consumption is only 11% that of urban consumption. In Fujian it is less unequal at 53%.

HEILONGJIANG
JILIN
LIAONING
XINJIANG AR
INNER MONGOLIA AR
GANSU
BEIJING
TIANJIN
HEBEI
NINGXIA AR
QINGHAI
SHANXI
SHANDONG
SHAANXI
HENAN
JIANGSU
ANHUI
SHANGHAI
TIBET AR
HUBEI
SICHUAN
CHONGQING
ZHEJIANG
JIANGXI
HUNAN
GUIZHOU
FUJIAN
YUNNAN
TAIWAN
GUANGXI AR
GUANGDONG
MACAU SAR
HONG KONG SAR
HAINAN

According to the World Bank, the number of people in poverty declined from 490 million in 1981 to 88 million in 2002. But in 2003 China admitted that poverty had risen for the first time since 1978.

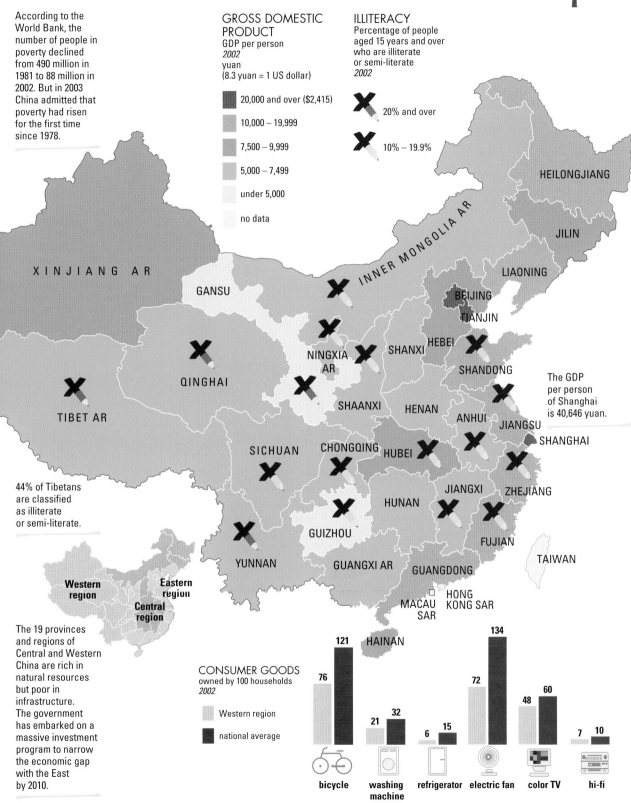

GROSS DOMESTIC PRODUCT
GDP per person
2002
yuan
(8.3 yuan = 1 US dollar)

- 20,000 and over ($2,415)
- 10,000 – 19,999
- 7,500 – 9,999
- 5,000 – 7,499
- under 5,000
- no data

ILLITERACY
Percentage of people aged 15 years and over who are illiterate or semi-literate
2002

- 20% and over
- 10% – 19.9%

The GDP per person of Shanghai is 40,646 yuan.

44% of Tibetans are classified as illiterate or semi-literate.

Western region

Eastern region

Central region

The 19 provinces and regions of Central and Western China are rich in natural resources but poor in infrastructure. The government has embarked on a massive investment program to narrow the economic gap with the East by 2010.

CONSUMER GOODS
owned by 100 households
2002

- Western region
- national average

	bicycle	washing machine	refrigerator	electric fan	color TV	hi-fi
Western region	76	21	6	72	48	7
national average	121	32	15	134	60	10

SEEING NO OX AS A WHOLE

Reliable data on employment and industrial output in China is notoriously difficult to assess, but changes over time show a clear picture. The private sector is growing and the state-owned enterprises are shrinking, although they remain a significant part of the economy.

Creating jobs is, and will remain, a daunting task, with at least 8 million new jobs a year needed to employ entrants to the labor market. Meanwhile, unpaid wages in China are estimated to reach US$15 billion, mainly in the construction sector.

Social scientists have turned their attention to the class structure. The dictum "Two classes, one stratum" no longer holds, and workers and peasants alike are relegated to near the bottom of the pile.

A study by the Chinese Academy of Social Sciences found that the middle class (defined as those with household assets over $18,000) had grown from 15 percent in 1999 to 19 percent in 2003.

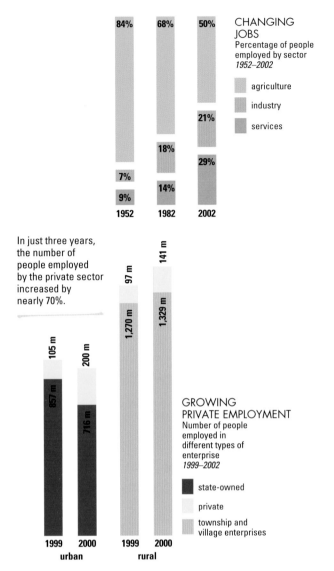

CHANGING JOBS
Percentage of people employed by sector
1952–2002

- agriculture
- industry
- services

	1952	1982	2002
agriculture	84%	68%	50%
industry	7%	18%	21%
services	9%	14%	29%

In just three years, the number of people employed by the private sector increased by nearly 70%.

GROWING PRIVATE EMPLOYMENT
Number of people employed in different types of enterprise
1999–2002

- state-owned
- private
- township and village enterprises

urban: 1999 (857 m, 105 m), 2000 (716 m, 200 m)
rural: 1999 (1,270 m, 97 m), 2000 (1,329 m, 141 m)

PRIVATE EMPLOYMENT
Sectoral share of privately funded and self-employed workforce
2002

- agriculture
- industry
- services

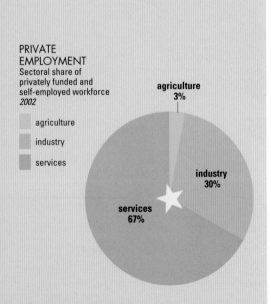

agriculture 3%
industry 30%
services 67%

CLASS STRUCTURE
By status as adapted by the Chinese Academy of Social Sciences
2004

1	State and society leaders
2	CEOs
3	Entrepreneurs
4	Professionals
5	Clerks
6	Shopkeepers
7	Commercial services staff
8	Industrial workers
9	Agricultural workers
10	Non-employed, unemployed, and semi-employed

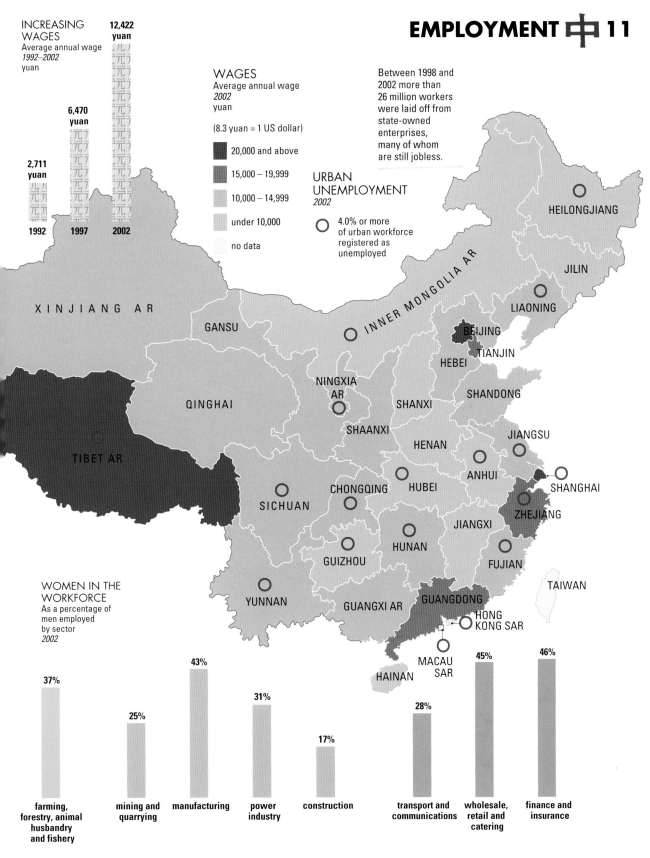

INCREASING WAGES
Average annual wage
1992–2002
yuan

2,711 yuan — **1992**
6,470 yuan — **1997**
12,422 yuan — **2002**

WAGES
Average annual wage
2002
yuan

(8.3 yuan = 1 US dollar)

- 20,000 and above
- 15,000 – 19,999
- 10,000 – 14,999
- under 10,000
- no data

URBAN UNEMPLOYMENT
2002

○ 4.0% or more of urban workforce registered as unemployed

Between 1998 and 2002 more than 26 million workers were laid off from state-owned enterprises, many of whom are still jobless.

Map labels: XINJIANG AR, GANSU, INNER MONGOLIA AR, HEILONGJIANG, JILIN, LIAONING, BEIJING, TIANJIN, HEBEI, SHANDONG, NINGXIA AR, SHANXI, SHAANXI, QINGHAI, HENAN, JIANGSU, TIBET AR, ANHUI, SHANGHAI, CHONGQING, HUBEI, ZHEJIANG, SICHUAN, JIANGXI, HUNAN, FUJIAN, GUIZHOU, TAIWAN, YUNNAN, GUANGXI AR, GUANGDONG, HONG KONG SAR, MACAU SAR, HAINAN

WOMEN IN THE WORKFORCE
As a percentage of men employed by sector
2002

Sector	Percentage
farming, forestry, animal husbandry and fishery	37%
mining and quarrying	25%
manufacturing	43%
power industry	31%
construction	17%
transport and communications	28%
wholesale, retail and catering	45%
finance and insurance	46%

THE THUNDER IS HUGE BUT THE RAINDROPS ARE MANY

Agriculture, once the ideological centerpiece of Maoist China, provides a declining share of China's GDP. While this is not out of line with trends in other industrializing countries, ensuring a continuing and adequate food supply for China's massive population presents the Chinese government with a particularly daunting logistical challenge.

Government investment has targeted agricultural modernization in an attempt to revive rural confidence and productivity. Animal husbandry (mainly factory farming), and production of crops vital to the food-processing industries are booming as a result.

The amount of grain produced has, however, declined in recent years, with falling grain prices a disincentive for farmers. In 2004 the government announced plans to improve productivity in the key grain-producing areas. Measures included slashing agricultural taxes and providing incentives to improve irrigation. Initial figures indicated that they were having the desired effect.

EMPLOYMENT
In agriculture as a percentage of total employed
2002

- 50% China ★
- 49% Thailand
- 47% Pakistan
- 45% Indonesia
- 42% Romania
- 38% Philippines
- 19% Poland
- 11% South Korea
- 5% Australia, Japan
- 3% USA
- 2% UK
- 1% France

AGRICULTURAL PRODUCTION
As a percentage of GDP
2002

- 38%
- 20% – 25%
- 15% – 19%
- 10% – 14%
- under 10%
- no data

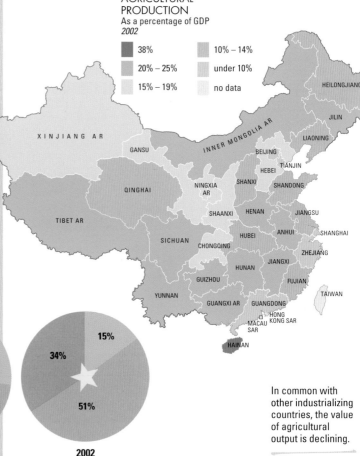

XINJIANG AR, GANSU, QINGHAI, TIBET AR, INNER MONGOLIA AR, HEILONGJIANG, JILIN, LIAONING, BEIJING, TIANJIN, HEBEI, SHANXI, SHANDONG, NINGXIA AR, SHAANXI, HENAN, JIANGSU, SICHUAN, CHONGQING, HUBEI, ANHUI, SHANGHAI, ZHEJIANG, HUNAN, JIANGXI, GUIZHOU, FUJIAN, YUNNAN, GUANGXI AR, GUANGDONG, HONG KONG SAR, MACAU SAR, HAINAN, TAIWAN

OUTPUT
As a percentage of GDP
1980–2002

- agriculture
- industry
- services

1980
- 21%
- 30%
- 49%

1990
- 31%
- 27%
- 43%

2002
- 15%
- 34%
- 51%

In common with other industrializing countries, the value of agricultural output is declining.

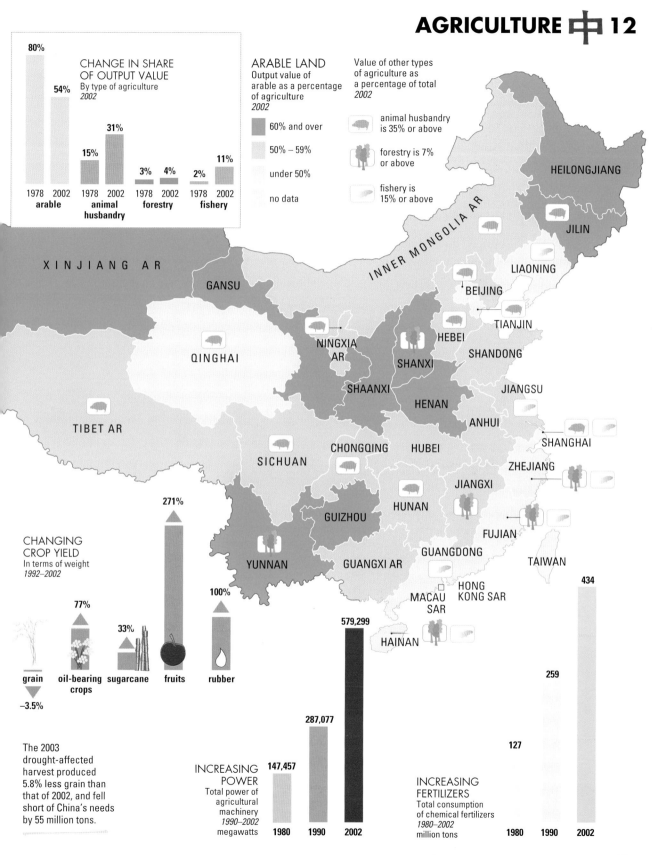

CHANGE IN SHARE OF OUTPUT VALUE
By type of agriculture
2002

- 80% 1978 / 54% 2002 — **arable**
- 15% 1978 / 31% 2002 — **animal husbandry**
- 3% 1978 / 4% 2002 — **forestry**
- 2% 1978 / 11% 2002 — **fishery**

ARABLE LAND
Output value of arable as a percentage of agriculture
2002

- 60% and over
- 50% – 59%
- under 50%
- no data

Value of other types of agriculture as a percentage of total
2002

- animal husbandry is 35% or above
- forestry is 7% or above
- fishery is 15% or above

Map labels

XINJIANG AR
GANSU
QINGHAI
TIBET AR
SICHUAN
YUNNAN
INNER MONGOLIA AR
NINGXIA AR
SHAANXI
CHONGQING
GUIZHOU
GUANGXI AR
HEILONGJIANG
JILIN
LIAONING
BEIJING
TIANJIN
HEBEI
SHANXI
SHANDONG
HENAN
JIANGSU
ANHUI
SHANGHAI
HUBEI
ZHEJIANG
JIANGXI
HUNAN
FUJIAN
GUANGDONG
TAIWAN
HONG KONG SAR
MACAU SAR
HAINAN

CHANGING CROP YIELD
In terms of weight
1992–2002

- grain **−3.5%**
- oil-bearing crops **77%**
- sugarcane **33%**
- fruits **271%**
- rubber **100%**

The 2003 drought-affected harvest produced 5.8% less grain than that of 2002, and fell short of China's needs by 55 million tons.

INCREASING POWER
Total power of agricultural machinery
1990–2002
megawatts

- **1980** 147,457
- **1990** 287,077
- **2002** 579,299

INCREASING FERTILIZERS
Total consumption of chemical fertilizers
1980–2002
million tons

- **1980** 127
- **1990** 259
- **2002** 434

LIKE A SILKWORM EATING THROUGH A LEAF

China's industrial output is booming, producing goods for export, and for the growing internal market. The output of all kinds of enterprises – state-owned, share-holding, and those funded from the Special Administrative Regions, Taiwan, or further afield – is increasing in value.

By the end of 2004, China was expected to be the third largest car manufacturer in the world, behind the USA and Japan.

State-owned industrial enterprises are at the top of the government's reform agenda, but the prospect of adding to urban unemployment makes them cautious.

EMPLOYMENT
In industry as a percentage of total employed
2002

- 32% Poland
- 31% Japan
- 28% South Korea
- 25% France, UK
- 23% USA
- 22% Australia
- 21% China ★
- 18% Thailand
- 17% Pakistan
- 16% Indonesia
- 15% Philippines

INDUSTRIAL PRODUCTION
As a percentage of GDP
2002

- ■ 50% and over
- ■ 40% – 49%
- ■ under 40%
- ▨ no data

China's industrial production, after waning during the 1980s, regained its position of providing 50% of GDP.

INDUSTRIAL EMPLOYMENT
By type of industry
2002

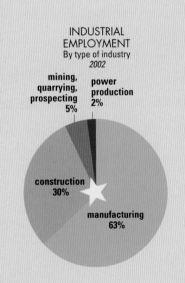

- mining, quarrying, prospecting 5%
- power production 2%
- construction 30%
- manufacturing 63%

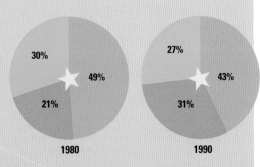

1980
- 30%
- 49%
- 21%

1990
- 27%
- 43%
- 31%

2002
- 15%
- 51%
- 34%

TRENDS IN INDUSTRIAL PRODUCTION
As a percentage of GDP
1980–2002

- ■ industry
- ■ services
- ■ agriculture

TYPE OF INDUSTRY
1998–2002

57% 1998 heavy industry
61% 2002 heavy industry
43% 1998 light industry
39% 2002 light industry

INDUSTRIAL ENTERPRISES
Number
2002

- 20,000 and over
- 10,000 – 13,499
- 5,000 – 9,999
- 1,000 – 4,999
- under 1,000
- no data

heavy industry (mining, smelting) is 70% or more of total

light industry (consumer goods and food processing) is 50% or more of total

Map labels

HEILONGJIANG
JILIN
LIAONING
INNER MONGOLIA AR
BEIJING
TIANJIN
HEBEI
SHANDONG
XINJIANG AR
GANSU
NINGXIA AR
SHANXI
SHAANXI
HENAN
JIANGSU
ANHUI
SHANGHAI
ZHEJIANG
QINGHAI
TIBET AR
SICHUAN
CHONGQING
HUBEI
JIANGXI
HUNAN
FUJIAN
TAIWAN
GUIZHOU
YUNNAN
GUANGXI AR
GUANGDONG
MACAU SAR
HONG KONG SAR
HAINAN

GROWTH IN DIFFERENT PRODUCTS
1990–2002
percentages

- 537% motor vehicles
- 501% plastics
- 399% color TVs
- 247% beer
- 245% refrigerators
- 170% paper
- 116% steel
- 59% sugar

CHANGING PUBLIC PRIVATE SPLIT
Value of output of enterprises by type of ownership
1998–2002
billion yuan

	1998	2002
state-owned	3,362	4,517
collective-owned	1,318	961
share-holding	433	1,412
funded from HK, Macau SAR, Taiwan	830	1,367
foreign-funded	846	1,879

IT DOESN'T MATTER
WHAT COLOR THE CAT IS,
SO LONG AS IT CATCHES
THE MICE

China's service sector, although growing, is still the junior partner when compared with industry. It lags behind those of the major industrialized countries in terms of employment, and accounts for only 34 percent of GDP. Certain services are booming, however, including real estate management, and enterprises, such as hotels and restaurants, associated with tourism.

Investment in information technology is vital to the creation of a modern service industry. Chinese government investment does not compare favorably with that of countries at a similar stage of economic development.

China's biggest challenge is to expand its services sector, with support from the IT sector, whilst maintaining its strong manufacturing base.

EMPLOYMENT
In services as a percentage of total employed
2002

75% USA
74% France
73% Australia, UK

63% Japan
61% South Korea

49% Malaysia
46% Philippines

39% Indonesia

36% Pakistan

33% Thailand

29% China ★

SERVICE EMPLOYMENT
By type of service
2002

wholesale, retail and catering **27%**

others, including tourism, hospitality, IT **34%**

transport, post and telecoms **11%**

scientific research **1%**

real estate 1%

finance and insurance 2%

healthcare and welfare **3%**

government and party agencies **6%**

social services **6%**

education, film, radio and TV **9%**

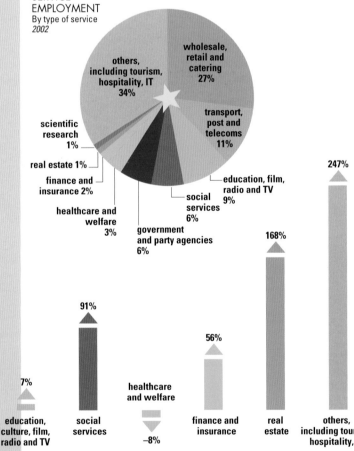

CHANGING EMPLOYMENT
Percentage change in number employed
1990–2002

75%
wholesale, retail trade and catering

33%
transport, post and telecoms

7%
education, culture, film, radio and TV

91%
social services

–8%
healthcare and welfare

56%
finance and insurance

168%
real estate

247%
others, including tourism, hospitality, IT

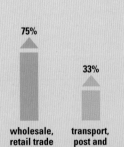

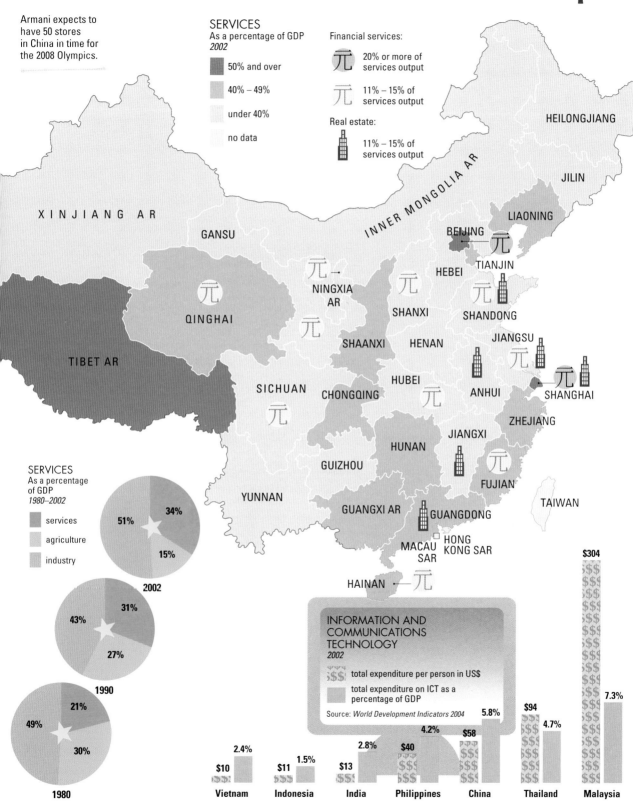

Armani expects to have 50 stores in China in time for the 2008 Olympics.

SERVICES
As a percentage of GDP
2002

- 50% and over
- 40% – 49%
- under 40%
- no data

Financial services:

元 20% or more of services output

元 11% – 15% of services output

Real estate:

11% – 15% of services output

HEILONGJIANG

JILIN

XINJIANG AR

GANSU

INNER MONGOLIA AR

LIAONING

BEIJING

TIANJIN

HEBEI

SHANDONG

NINGXIA AR

SHANXI

QINGHAI

SHAANXI

HENAN

JIANGSU

TIBET AR

SICHUAN

CHONGQING

HUBEI

ANHUI

SHANGHAI

ZHEJIANG

JIANGXI

HUNAN

FUJIAN

GUIZHOU

YUNNAN

GUANGXI AR

GUANGDONG

TAIWAN

MACAU SAR

HONG KONG SAR

HAINAN

SERVICES
As a percentage of GDP
1980–2002

- services
- agriculture
- industry

2002
34%
15%
51%

1990
31%
27%
43%

1980
21%
30%
49%

INFORMATION AND COMMUNICATIONS TECHNOLOGY
2002

$$$ total expenditure per person in US$

total expenditure on ICT as a percentage of GDP

Source: *World Development Indicators 2004*

	Vietnam	Indonesia	India	Philippines	China	Thailand	Malaysia
% of GDP	2.4%	1.5%	2.8%	4.2%	5.8%	4.7%	7.3%
per person	$10	$11	$13	$40	$58	$94	$304

THE GRASS STIRS
AS THE WIND BLOWS

At the peak of the Cultural Revolution between 1966 and 1968, millions of young people traveled the country to spread continuing revolution as a doctrine of transformation. Now the travelers are Chinese tourists, most of them enriched by trade in the new market economy.

In 2008, the focus will be on international arrivals, whose foreign dollars will finance the huge costs of reinventing Beijing for China's first Olympic Games. The city is building 22 new sports venues and renovating five existing spaces to the tune of US$251.53 million. Beijing's infrastructure is being upgraded to enable extensive wireless laptop access and good sanitation for the assumed millions who will come to applaud China's medal tally.

By the end of 2010 it will all have happened again in Shanghai, this time for World Expo.

The Beijing Olympic Committee is planning an international food street for the Olympic tourists. In 2004, while Guangzhou had 36,000 western-style restaurants, there were 1,600 in Shanghai and only 600 in Beijing.

In 2003 domestic tourists brought in 70.6 billion yuan. International tourists brought in US$ 1.9 million. (8.3 yuan = 1 US dollar)

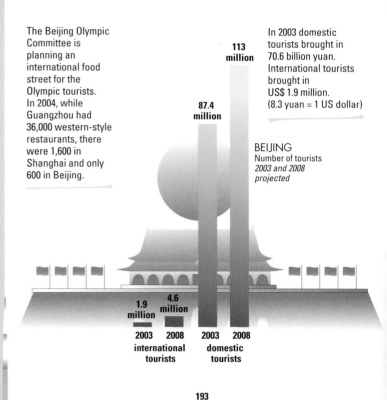

113 million

87.4 million

BEIJING
Number of tourists
2003 and 2008 projected

1.9 million **4.6 million**

2003	2008	2003	2008
international tourists		domestic tourists	

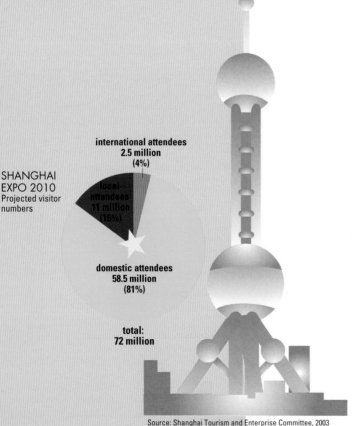

SHANGHAI EXPO 2010
Projected visitor numbers

international attendees
2.5 million (4%)

local attendees
11 million (15%)

domestic attendees
58.5 million (81%)

total:
72 million

193

156.8

109.8

2002	2005	2010

TOURIST EARNINGS
Projected earnings for Shanghai from domestic tourism
2002, 2005 and 2010 projected
yuan billion

$4.4

$3.2

$2.3

2002	2005	2010

FOREIGN EXCHANGE
Projected earnings for Shanghai from foreign tourism
2002, 2005 and 2010 projected
US$ billion

Source: Shanghai Tourism and Enterprise Committee, 2003

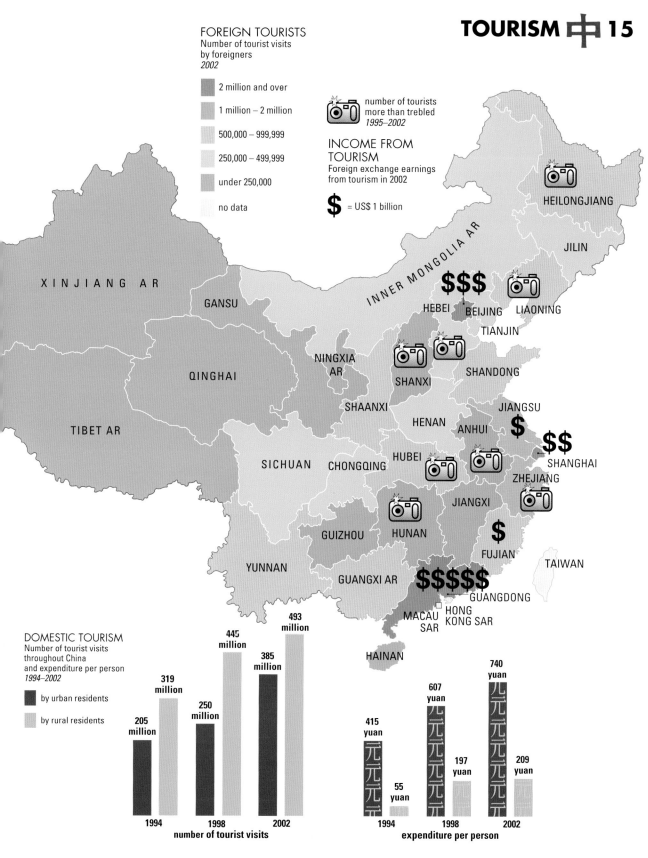

FOREIGN TOURISTS
Number of tourist visits by foreigners
2002

- 2 million and over
- 1 million – 2 million
- 500,000 – 999,999
- 250,000 – 499,999
- under 250,000
- no data

number of tourists more than trebled *1995–2002*

INCOME FROM TOURISM
Foreign exchange earnings from tourism in 2002

$ = US$ 1 billion

HEILONGJIANG

JILIN

INNER MONGOLIA AR

XINJIANG AR

GANSU

$$$

HEBEI   BEIJING   LIAONING

TIANJIN

NINGXIA AR

SHANXI   SHANDONG

QINGHAI

SHAANXI

JIANGSU

$

$$
SHANGHAI

TIBET AR

HENAN   ANHUI

SICHUAN   CHONGQING

HUBEI

ZHEJIANG

JIANGXI

GUIZHOU   HUNAN

$
FUJIAN

TAIWAN

YUNNAN

GUANGXI AR

$$$$$
GUANGDONG

MACAU SAR   HONG KONG SAR

HAINAN

DOMESTIC TOURISM
Number of tourist visits throughout China and expenditure per person
1994–2002

- by urban residents
- by rural residents

number of tourist visits

Year	by urban residents	by rural residents
1994	205 million	319 million
1998	250 million	445 million
2002	385 million	493 million

expenditure per person

Year	by urban residents	by rural residents
1994	415 yuan	55 yuan
1998	607 yuan	197 yuan
2002	740 yuan	209 yuan

JOURNEY TO THE WEST: FOUR WHEELS GOOD, TWO WHEELS BAD

The growth of motorized traffic in China's cities – especially in those along the eastern seaboard – is both indicative of the drive towards rapid modernization, and of the dangers that development can produce.

Up to 90 percent of emissions in major cities can be attributed to motor vehicles, car accidents are common, and luxury car ownership is on the rise. Many drivers and pedestrians are as yet unused to the etiquettes of safety in motorized environments – and the notion of right of way is an ongoing dispute, which frequently results in road rage and fatalities.

The high demand for cars in China has encouraged Western companies to pledge an investment of US$13 billion on China's car-manufacturing facilities, thereby tripling the country's output.

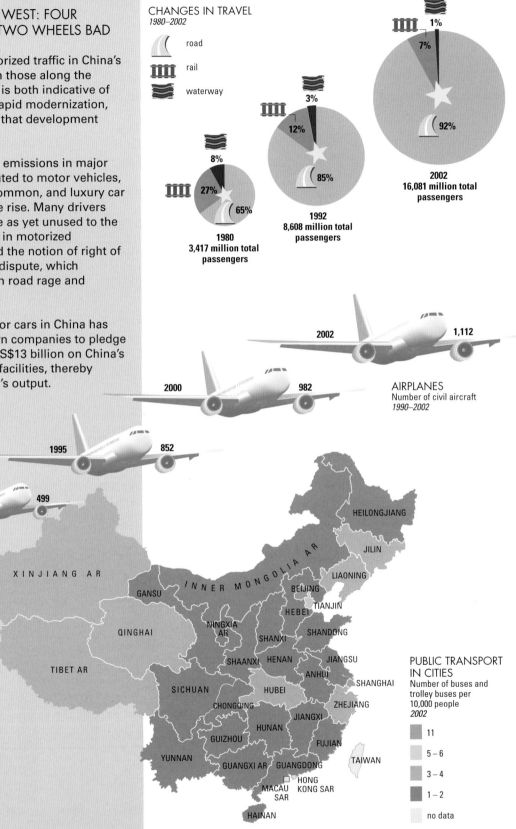

CHANGES IN TRAVEL
1980–2002

road
rail
waterway

1980
3,417 million total passengers
65%
27%
8%

1992
8,608 million total passengers
85%
12%
3%

2002
16,081 million total passengers
92%
7%
1%

AIRPLANES
Number of civil aircraft
1990–2002

2002 — 1,112
2000 — 982
1995 — 852
1990 — 499

PUBLIC TRANSPORT IN CITIES
Number of buses and trolley buses per 10,000 people
2002

11
5 – 6
3 – 4
1 – 2
no data

HEILONGJIANG
JILIN
LIAONING
XINJIANG AR
INNER MONGOLIA AR
BEIJING
GANSU
TIANJIN
HEBEI
NINGXIA AR
SHANDONG
SHANXI
QINGHAI
SHAANXI
HENAN
JIANGSU
TIBET AR
ANHUI
SHANGHAI
SICHUAN
HUBEI
ZHEJIANG
CHONGQING
JIANGXI
HUNAN
FUJIAN
GUIZHOU
YUNNAN
GUANGXI AR
GUANGDONG
TAIWAN
HONG KONG SAR
MACAU SAR
HAINAN

TRAFFIC ACCIDENTS
Number a year
2002

- 40,000 and over
- 20,000 – 39,999
- 10,000 – 19,999
- under 10,000
- no data

600 people a day
were killed in traffic
accidents in 2004.

HEILONGJIANG

JILIN

XINJIANG AR

GANSU

INNER MONGOLIA AR

LIAONING

BEIJING

NINGXIA
AR

SHANXI

HEBEI

TIANJIN

QINGHAI

SHANDONG

SHAANXI

HENAN

JIANGSU

TIBET AR

ANHUI

SHANGHAI

SICHUAN

CHONGQING

HUBEI

ZHEJIANG

HUNAN

JIANGXI

GUIZHOU

FUJIAN

YUNNAN

TAIWAN

GUANGXI AR

GUANGDONG

HONG
KONG SAR

MACAU
SAR

HAINAN

HEILONGJIANG
21

XINJIANG AR
31

JILIN
18

GANSU
18

INNER MONGOLIA AR
48

21
LIAONING

QINGHAI
26

BEIJING
56

TIBET AR
54

**NINGXIA
AR**
34

SHANXI
30

HEBEI
37

TIANJIN
63

SHANDONG
27

SHAANXI
15

HENAN
20

JIANGSU
14

SICHUAN
24

CHONGQING
15

HUBEI
16

ANHUI
14

SHANGHAI
2

ZHEJIANG
46

GUIZHOU
20

HUNAN
24

JIANGXI
10

FUJIAN
32

YUNNAN
41

TAIWAN

GUANGXI AR
14

GUANGDONG
87

**MACAU
SAR**

**HONG
KONG SAR**

HAINAN
26

PRIVATE TRANSPORT
Number of passenger
vehicles owned
per 10,000 people
2002

- 200 and over
- 100 – 199
- 50 – 99
- 25 – 49
- 10 – 24
- no data

Number of trucks
owned per 10,000 people
2002

- 50 and over
- 25 – 49
- under 25

China's total road
mileage is now the
third largest in the
world.

By 2020, there are
expected to be
seven times the
number of cars on
China's roads than
in 2004.

A HEAVY WEIGHT HANGS BY A HAIR

China is the world's leading coal producer. Coal supplies two-thirds of China's energy needs. The abundance of coal, much of it low-grade, has environmental and health costs, and the government is investing in research into cost-effective coal substitutes.

Despite China's progress in energy production, it cannot keep up with the increasing demand for electricity from its rapidly expanding economy. Since mid-2002 there has been a serious power shortfall, which is undermining the confidence of overseas investors. Optimistic predictions indicate that the power supply will once again meet demand by 2006, but oil and natural gas are under-developed as sources of power, and hydro-power is under-used. New nuclear power stations are being commissioned, but will not come on line for some years.

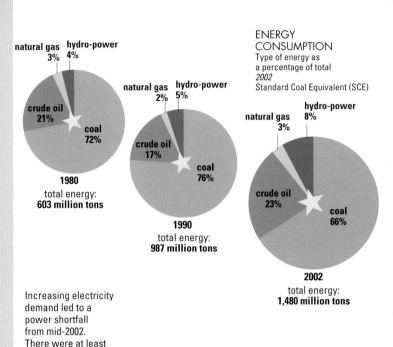

ENERGY CONSUMPTION
Type of energy as a percentage of total
2002
Standard Coal Equivalent (SCE)

natural gas 3%
hydro-power 4%
crude oil 21%
coal 72%
1980
total energy: **603 million tons**

natural gas 2%
hydro-power 5%
crude oil 17%
coal 76%
1990
total energy: **987 million tons**

hydro-power 8%
natural gas 3%
crude oil 23%
coal 66%
2002
total energy: **1,480 million tons**

Increasing electricity demand led to a power shortfall from mid-2002. There were at least 175,000 power outages in 2004, mainly in the south and east. Sony and Volkswagen were forced to cut back on production.

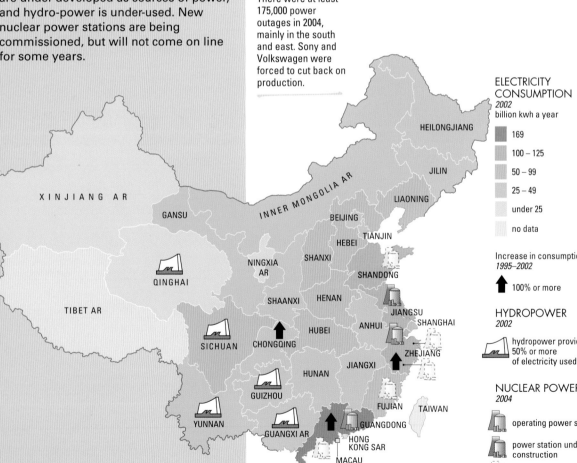

ELECTRICITY CONSUMPTION
2002
billion kwh a year

- 169
- 100 – 125
- 50 – 99
- 25 – 49
- under 25
- no data

Increase in consumption
1995–2002

↑ 100% or more

HYDROPOWER
2002

hydropower provides 50% or more of electricity used

NUCLEAR POWER
2004

🏭 operating power stations

🏭 power station under construction

🏭 proposed power station

Source: World Nuclear Association
www.world-nuclear.org

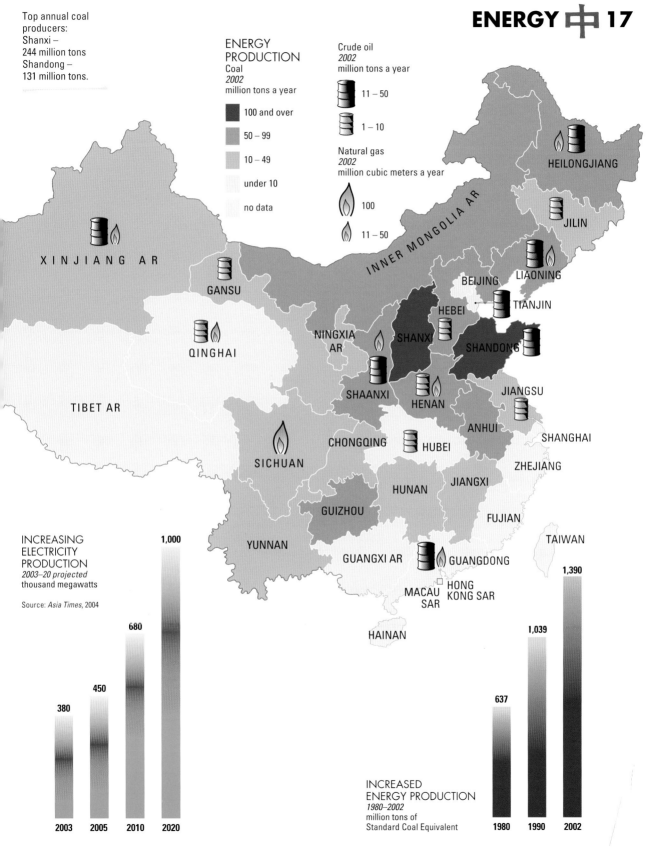

Top annual coal
producers:
Shanxi –
244 million tons
Shandong –
131 million tons.

ENERGY
PRODUCTION
Coal
2002
million tons a year

100 and over

50 – 99

10 – 49

under 10

no data

Crude oil
2002
million tons a year

11 – 50

1 – 10

Natural gas
2002
million cubic meters a year

100

11 – 50

HEILONGJIANG

JILIN

INNER MONGOLIA AR

LIAONING

XINJIANG AR

GANSU

BEIJING

HEBEI

TIANJIN

NINGXIA
AR

SHANXI

SHANDONG

QINGHAI

SHAANXI

JIANGSU

TIBET AR

HENAN

ANHUI

SHANGHAI

CHONGQING

HUBEI

ZHEJIANG

SICHUAN

HUNAN

JIANGXI

GUIZHOU

FUJIAN

YUNNAN

TAIWAN

GUANGXI AR

GUANGDONG

MACAU
SAR

HONG
KONG SAR

HAINAN

INCREASING
ELECTRICITY
PRODUCTION
2003–20 projected
thousand megawatts

Source: *Asia Times*, 2004

1,000

680

450

380

2003 | 2005 | 2010 | 2020

INCREASED
ENERGY PRODUCTION
1980–2002
million tons of
Standard Coal Equivalent

1,390

1,039

637

1980 | 1990 | 2002

Part Four

THE PARTY-STATE

CHINA'S POLITICAL REFORMS have not kept pace with the rapid restructuring of the economy. The Communist Party retains its monopoly of power, including its control of the military, and sustains an authoritarian state, with its repression and civil rights abuses. Nonetheless, important changes are shaping China's political development.

Firstly, the gap between the economic reforms and the political reforms are deliberate policy. Moving away from a planned economy towards a market-led economy is reforming a revolution. It is no wonder that the Party-state's ruling elite insists that political stability is equal in priority to the market reforms. This is reinforced by the break-up of the former Soviet Union. Political stability is also valued in its own right, given the history of modern China's turbulence and disorder.

Secondly, political reform has been making steady progress. The bureaucracy has not only been modernized but is also more professional in terms of the education, training and competence of its officials. The Party is also attempting to broaden its representation in line with Jiang Zemin's "Three Represents", which called for the admission of entrepreneurs to Party membership. But power remains firmly in the hands of a small elite. This is evidenced by the unopposed succession to office of the Fourth Generation of Party-state leaders, their consolidation of power with the resignation of the former Party-state chief, Jiang Zemin, as Chairman of the Central Military Commission, and his replacement by the current Party Secretary and President, Hu Jintao.

Experiments have also been taking place for electing Party branch secretaries at the grass-roots level. Dramatic changes are taking in the villages and the urban neighborhoods. Villages have been electing leaders and holding them to account

There were 58,000 major incidents of social unrest in 2003 – six times the number in 1993.

since 1988. Five rounds of nationwide elections have taken place and a sixth is to begin this year. In the more prosperous villages, leaders exercise power over property rights, land use and the distribution of welfare funds. Experiments in direct elections are taking place at the next levels up: the towns and villages. In the cities, the new Community Residents' Committees have greater responsibilities than they did in the 1950s, and are electing younger, better-educated leaders to exercise them. While this grass-roots political activity bodes well for democratic practice, much will depend on the extent of Party control.

Thirdly, with the growth of the market-led economy, producers and consumers have formed organizations to represent, promote and defend their interests. In China, these are mainly trade and professional organizations and they are obliged to register with the government for legal recognition. For example, in the northeast city, Shenyang, there is the Taxi Drivers Association, the Pharmacy Association and the Football Fans Association, among others. In addition, there are the Communist Party pillar or mass organizations, the unofficial and the underground organizations. Unofficial organizations are growing at such a rate that the Party-state is considering abolishing the registration system. Also growing in numbers are the often spontaneous and usually local protest movements, organized by peasants, workers, pensioners and ethnic minorities, for example. They do not necessarily threaten the regime but they do erode political stability.

Another view is that political movements, along with the organized interests, enrich the politics that promote civil society. Whether or not the development of civil society in an authoritarian Party-state is a harbinger of radical change, it is an important arena for political activity and conflict.

A TIGER WHOSE BUTTOCKS MAY NOT BE TOUCHED

The Chinese Communist Party, with a membership of 68 million in 2004, is the largest political party in the world.

In 2002 the Party adopted Jiang Zemin's "Three Represents" as a major theoretical contribution that called for the opening of the Party to entrepreneurs, intellectuals, and scientists. The 16th Party Congress also saw the succession to power of the Fourth Generation of Party leaders.

The eight small democratic parties do not provide an opposition to the Communist Party. Any challenge to the Communist Party's monopoly of power is likely to come from within its own ranks rather than from without.

INCREASE IN PARTY MEMBERSHIP
Number of members of the Chinese Communist Party

Source: Saich, 2004

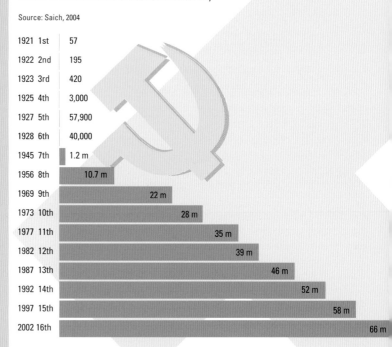

1921	1st	57
1922	2nd	195
1923	3rd	420
1925	4th	3,000
1927	5th	57,900
1928	6th	40,000
1945	7th	1.2 m
1956	8th	10.7 m
1969	9th	22 m
1973	10th	28 m
1977	11th	35 m
1982	12th	39 m
1987	13th	46 m
1992	14th	52 m
1997	15th	58 m
2002	16th	66 m

DEMOCRATIC PARTIES
2004
Under the leadership
of the Communist
Party of China

Source: State Council Information Office

China Democratic League
September 3rd Study Society
high-level academics in social and natural sciences

Chinese Peasants' and Workers' Democratic Party
China Association for Promoting Democracy
professionals, doctors

China Democratic National Construction Association
business people

Revolutionary Committee of the Guomindang
Taiwan Democratic Self-Government League
Party for Public Interest
people with Taiwan connections and overseas Chinese

INFORMAL POWER
Guanxi: using contacts,
friends and family to get things done

POWER STRUCTURE
Standing Committee
of the Politburo
2004

Hu Jintao
General Secretary of Communist Party 2002–
President of People's Republic of China 2003–
Chairman of Central Military Commission 2004–
born 1942, Anhui • water conservancy engineer, Qinghua University • joined CCP 1964, Vice-President of People's Republic of China 1998–2003, President of Central Party School 1993–2002, Party Secretary of Guizhou 1985–88 and Tibet 1988–92 • entered Politburo and Standing Committee 1992

Zeng Qinghong
Vice-President of People's Republic of China 2003–
Secretary of Central Secretariat of CCP 1997–
President of Central Party School 2002–
born 1939, Jiangxi • automation engineer, Beijing Institute of Industry • joined CCP 1960 • Chief of Department of Organization of the Party 1999–2002 • Vice-Director of Office of Central Committee 1989–93 and Director 1993–99 • Deputy Party Secretary of Shanghai 1986–89 • entered Politburo as alternate member 1997, full member 2002

Wen Jiabao
Premier of the State Council 2003–
born 1942, Tianjin • geological engineer, Beijing Institute of Geology • joined CCP 1965 • Vice-Premier of State Council 1998–2003 • Secretary of Central Secretariat of Party 1992–2002 • Director of Office of Party Central Committee 1986–93 • entered Politburo as alternate member 1992, full member 1997–

Wu Bangguo
Chair of Standing Committee of National People's Congress 2003–
born 1941, Anhui • electrical engineer, Qinghua University • joined CCP 1964 • Vice-Premier of State Council 1995–2003 • Secretary of Central Secretariat 1994–97 • Deputy Party Secretary of Shanghai 1985–1991 and Secretary 1991–94 • entered Politburo 1992

Jia Qinglin
Chairman of Chinese People's Political Consultative Conference 2003–
born 1940, Hebei • electrical engineer, Hebei Institute of Engineering • Party Secretary of Fujian 1993–96 and Beijing 1997–2002 • Mayor of Beijing 1997–99 • entered Politburo 1997

Huang Ju
Vice-Premier of the State Council 2003–
born 1938, Zhejiang • electrical engineer, Qinghua University • joined CCP 1962 • Party Secretary of Shanghai 1994–2002 • Mayor of Shanghai 1991–94 • entered Politburo 1994

Wu Guanzheng
Secretary of Central Discipline Inspection Commission 2002–
born 1938, Jiangxi • dynamics engineer, Qinghua University • joined CCP 1962 • Party Secretary of Jiangxi 1995–97 and Shandong 1997–2002 • entered Politburo 1997

Li Changchun
Standing Committee member in charge of the ideological front 2002–
born 1944, Liaoning • electrical engineer, Harbin University of Industry • joined CCP 1965 • Party Secretary of Henan 1992–98 and Guangdong 1998–2002 • Governor of Henan 1991–92 • Deputy Secretary of Liaoning 1985–90 and Henan 1990–92 • entered Politburo 1997

Luo Gan
Secretary of Central Political and Legal Commission 2002–
born 1935, Shandong • metallurgical engineer, Freiburg Institute of Metallurgy, East Germany • joined CCP 1960 • Secretary of Central Secretariat 1997–2002 • State Councilor 1993–2003 • Chief Secretary of State Council 1988–97 • Minister of Labor 1988 • entered Politburo 1997

Politburo
plus 15 other full members (all at same level) and one alternate member

Central Military Commission
Chair: Hu Jintao • Vice-Chairs: Guo Boxiong, Cao Gangchuan, Xu Cai • six other members

Central Discipline Inspection Commission
Secretary: Wu Guanzheng
Seven Deputy Secretaries

Central Committee
198 members and 158 alternate members

National Party Congress meets every 5 years
The 16th Congress (2002): 2,114 delegates, 40 specially invited delegates

TO FIND A NEEDLE OR A PEARL IN THE BIG OCEAN

China's constitution states that the National People's Congress is the highest organ of state power. Relative to the Communist Party, the State Council, the Military Commission and the workings of informal politics, this is not the case. The pyramids represent a more realistic hierarchy.

Since 1978, however, the National People's Congress has been behaving more like a legislature familiar to Western eyes. It no longer simply rubber-stamps decisions taken elsewhere, and it has become a more professional organization with a support staff of over 3,000.

Political reform still lags behind economic reform. However, at the grass-roots level there has been a significant growth of elected village committees with decision-making powers. Elections are being extended to townships and to the reconstituted community residents' committees in the cities.

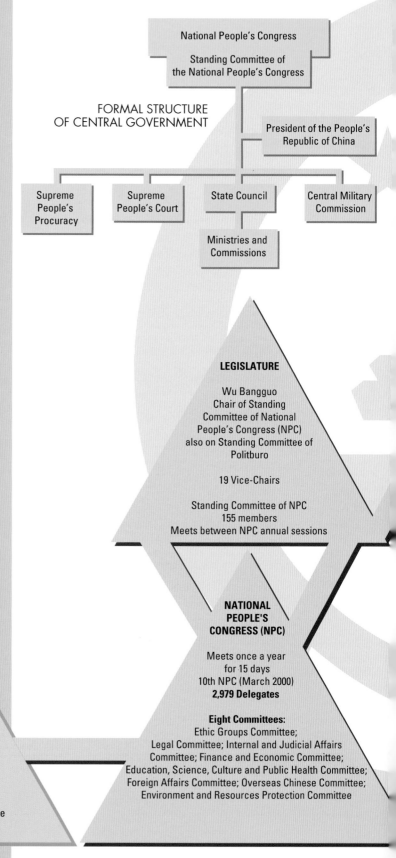

FORMAL STRUCTURE OF CENTRAL GOVERNMENT

National People's Congress

Standing Committee of the National People's Congress

President of the People's Republic of China

Supreme People's Procuracy

Supreme People's Court

State Council

Central Military Commission

Ministries and Commissions

LEGISLATURE

Wu Bangguo
Chair of Standing Committee of National People's Congress (NPC) also on Standing Committee of Politburo

19 Vice-Chairs

Standing Committee of NPC
155 members
Meets between NPC annual sessions

NATIONAL PEOPLE'S CONGRESS (NPC)

Meets once a year for 15 days
10th NPC (March 2000)
2,979 Delegates

Eight Committees:
Ethic Groups Committee; Legal Committee; Internal and Judicial Affairs Committee; Finance and Economic Committee; Education, Science, Culture and Public Health Committee; Foreign Affairs Committee; Overseas Chinese Committee; Environment and Resources Protection Committee

HONG KONG
Special Administrative Region (SAR)

Chief Executive
Tung Chee-hwa
recommended locally, appointed by Beijing central government 1997; reappointed 2002

Legislative Council
60 members elected 2004 for 4 years:
24 by proportional representation
30 indirectly by functional constituencies
6 appointed by 800-member election committee

EXECUTIVE

Hu Jintao
President
also General Secretary of
Communist Party

Zeng Qinghong
Vice-President
also on Standing Committee of Politburo

Wen Jiabao
Premier of
the State Council
also on Standing Committee of Politburo

State Council
Premier
4 Vice-Premiers
5 State Councilors
Secretary General
28 members at ministerial level
28 ministries and commissions,
51 offices, bureaux and institutions

POWER STRUCTURE OF
CENTRAL GOVERNMENT
2004

Source: People's Daily Online

MILITARY

Central Military
Commission (State)

Chair
Hu Jintao
also Chair of Communist Party
Central Military Commission

JUDICIARY
Supreme
People's Court
President
Xiao Yang
Procurator General
Jia Chunwang

Supreme People's
Procuratorate

**CHINESE
PEOPLE'S
POLITICAL
CONSULTATIVE
CONGRESS (CPPCC)**
over 2,000 delegates

Jia Qinglin
Chair, CPPCC
also on Standing Committee of Politburo

24 Vice-Chairs, Secretary General
299 Members of the Standing Committee

The CPPCC meets once a year, in conjunction
with the NPC

The 2,000-plus delegates include intellectuals, academics, business
people, technical experts, overseas Chinese, professionals,
democratic parties

SHARPENING THE WEAPONS AND FEEDING THE HORSES

The Communist Party controls the PLA through the Central Military Commission, chaired by Party-state boss Hu Jintao. It is a symbiotic relationship, however, for the Party-state relies ultimately on the PLA for maintaining domestic stability and for helping to project China as an international power, while the PLA is beholden to the Party-state for its large-scale modernization program.

The substantial increases in the military budget over the last 15 years testify to this support. The immediate focus for the military build-up, however, is only 160 km across the Taiwan Straits. Even so, this may have less to do with national security or power than with national pride.

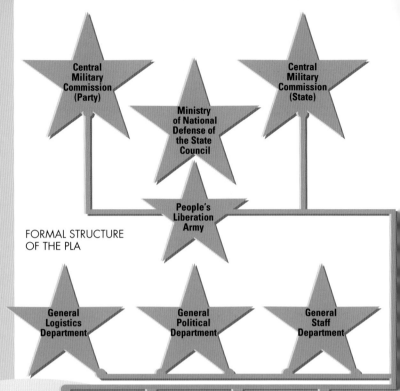

FORMAL STRUCTURE
OF THE PLA

ARMY DEPLOYMENT
Number of soldiers
by military region
2003

200,000 or more

under 200,000

Source: IISS, *Military Balance 2004–05*

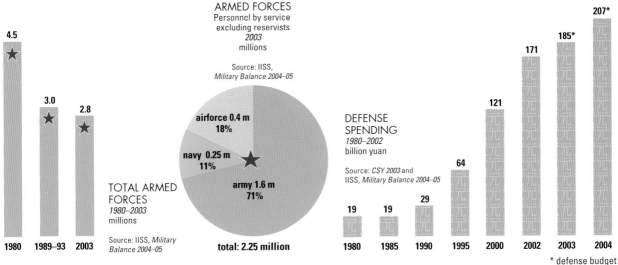

ARMED FORCES
Personnel by service
excluding reservists
2003
millions

Source: IISS,
Military Balance 2004–05

airforce 0.4 m
18%

navy 0.25 m
11%

army 1.6 m
71%

total: 2.25 million

TOTAL ARMED FORCES
1980–2003
millions

Source: IISS, *Military Balance 2004–05*

4.5
3.0
2.8

1980   1989–93   2003

DEFENSE SPENDING
1980–2002
billion yuan

Source: *CSY 2003* and
IISS, *Military Balance 2004–05*

19   19   29   64   121   171   185*   207*

1980   1985   1990   1995   2000   2002   2003   2004

* defense budget

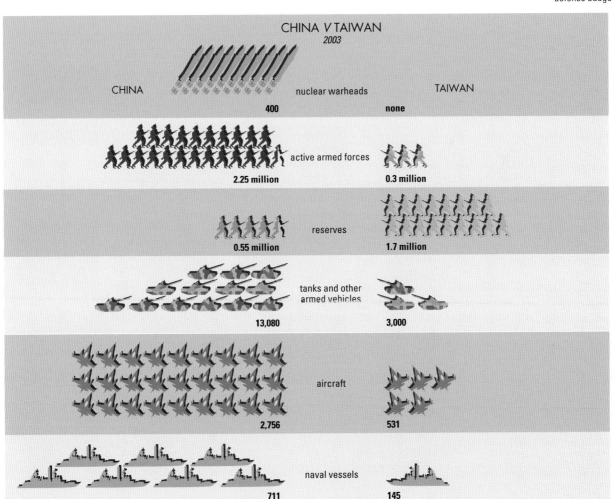

CHINA *V* TAIWAN
2003

CHINA   TAIWAN

nuclear warheads
400   none

active armed forces
2.25 million   0.3 million

reserves
0.55 million   1.7 million

tanks and other armed vehicles
13,080   3,000

aircraft
2,756   531

naval vessels
711   145

THE MOUNTAINS ARE HIGH AND THE EMPEROR IS FAR AWAY

Party and government structures run in parallel, extending from the center of power to the grass roots, where Village Committees and Community Residents' Committees are elected throughout China.

The coastal provinces take the lion's share of foreign direct investment. In 2002, this amounted to 89 percent of the total, increasing economic inequality between the regions. Such a massive injection of foreign capital gives these provinces a certain financial independence from central government.

Extra-budgetary revenue, sometimes known as the "second budget", is also an important source of local power and income. It takes the form of administrative fees and other charges levied on almost all aspects of life, including fees for every pig slaughtered, for animal inoculations, for school, and for permits to marry or to have children.

GOVERNMENT REVENUE AND EXPENDITURE
2004
billion yuan

local government
851
45%

central government
1,039
55%

total revenue: 1,890

local government
1,528
69%

central government
677
31%

total expenditure: 2,205

The gap in local government budgets between revenue and expenditure is made up by numerous fees and charges. Some 360 have been levied over the past decade.

ORGANIZATION OF COMMUNIST PARTY

Source: authors

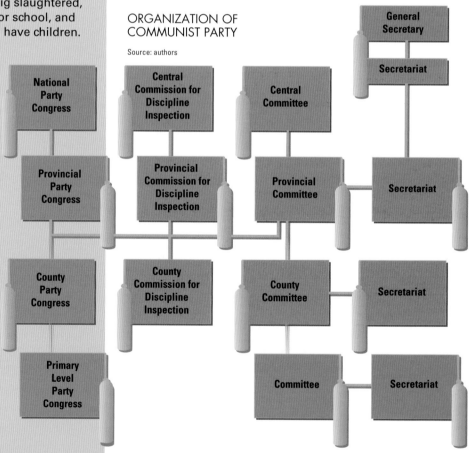

National Party Congress

Central Commission for Discipline Inspection

Central Committee

General Secretary

Secretariat

Provincial Party Congress

Provincial Commission for Discipline Inspection

Provincial Committee

Secretariat

County Party Congress

County Commission for Discipline Inspection

County Committee

Secretariat

Primary Level Party Congress

Committee

Secretariat

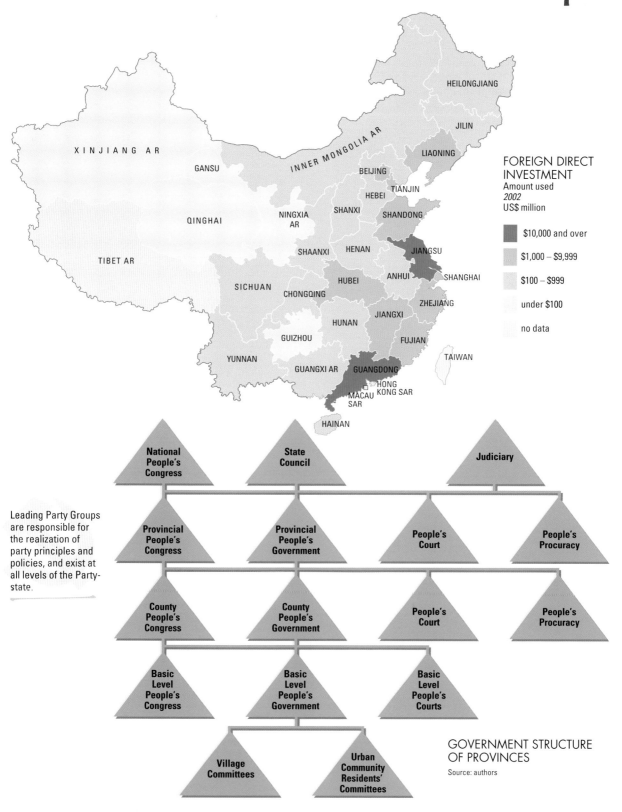

FOREIGN DIRECT
INVESTMENT
Amount used
2002
US$ million

$10,000 and over

$1,000 – $9,999

$100 – $999

under $100

no data

Leading Party Groups
are responsible for
the realization of
party principles and
policies, and exist at
all levels of the Party-
state.

National
People's
Congress

State
Council

Judiciary

Provincial
People's
Congress

Provincial
People's
Government

People's
Court

People's
Procuracy

County
People's
Congress

County
People's
Government

People's
Court

People's
Procuracy

Basic
Level
People's
Congress

Basic
Level
People's
Government

Basic
Level
People's
Courts

Village
Committees

Urban
Community
Residents'
Committees

GOVERNMENT STRUCTURE
OF PROVINCES

Source: authors

MOUNTAINS AND RIVERS ARE EASY TO MOVE BUT IT IS IMPOSSIBLE TO CHANGE THE NATURE OF A MAN

China in the era of economic reform has taken an important step towards legality, moving away from the Maoist principle of rule by persons to rule by law, and towards enshrining the rule of law in the Constitution.

Market and property relations require formal procedures and rules. While mediation remains an important means of settling domestic conflicts, economic disputes make up over one fifth of all cases coming before the courts. At the same time, *guanxi* – using informal contacts – is embedded in China's economic, political, and social life. Although *guanxi* may benefit those with personal ties, authority is eroded and enforcement is undermined.

Corruption flourishes in this environment. Despite the enactment of more than 2,000 anti-corruption laws and regulations between 1998 and 2003, there has been only a small increase in the number of corruption and bribery cases coming to court.

LITTLE ROOM FOR APPEAL
In 99% of trials, the accused is convicted. The appeals process offers little hope. Of the nearly 130,000 appeals accepted in 2002 only 4% have been resolved, and in only 0.5% of those cases was the original decision changed.

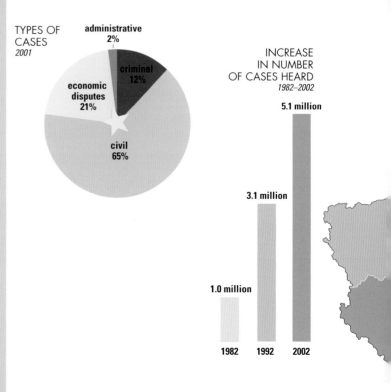

TYPES OF CASES
2001

- administrative 2%
- criminal 12%
- economic disputes 21%
- civil 65%

INCREASE IN NUMBER OF CASES HEARD
1982–2002

- 1.0 million — 1982
- 3.1 million — 1992
- 5.1 million — 2002

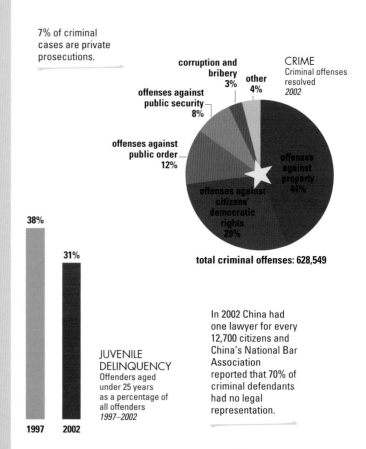

7% of criminal cases are private prosecutions.

CRIME
Criminal offenses resolved
2002

- corruption and bribery 3%
- other 4%
- offenses against public security 8%
- offenses against public order 12%
- offenses against citizens' democratic rights 29%
- offenses against property 44%

total criminal offenses: 628,549

JUVENILE DELINQUENCY
Offenders aged under 25 years as a percentage of all offenders
1997–2002

- 38% — 1997
- 31% — 2002

In 2002 China had one lawyer for every 12,700 citizens and China's National Bar Association reported that 70% of criminal defendants had no legal representation.

LAW ENFORCEMENT EXPENDITURE

Annual spending on public security and armed police per person
2002
yuan

- 300 and over
- 200 – 299
- 100 – 199
- 50 – 99
- under 50
- no data

XINJIANG AR

GANSU

QINGHAI

TIBET AR

SICHUAN

YUNNAN

INNER MONGOLIA AR

HEILONGJIANG

JILIN

LIAONING

BEIJING

TIANJIN

HEBEI

NINGXIA AR

SHANXI

SHANDONG

SHAANXI

HENAN

JIANGSU

ANHUI

SHANGHAI

HUBEI

CHONGQING

ZHEJIANG

JIANGXI

HUNAN

GUIZHOU

FUJIAN

TAIWAN

GUANGXI AR

GUANGDONG

HONG KONG SAR

MACAU SAR

HAINAN

MEDIATION

Types of civil dispute resolved by the People's Mediation Committees
2002

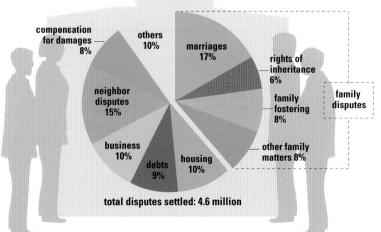

- compensation for damages 8%
- others 10%
- marriages 17%
- rights of inheritance 6%
- family fostering 8%
- neighbor disputes 15%
- business 10%
- debts 9%
- housing 10%
- other family matters 8%

family disputes

total disputes settled: 4.6 million

CONSTITUTION

modified by the National People's Congress

FUNDAMENTAL LAWS

enacted and modified by the National People's Congress

LAWS

enacted and modified by the Standing Committee of the National People's Congress

ADMINISTRATIVE RULES AND REGULATIONS

enacted and revised by the State Council

REGULATIONS

made by Ministries or Committees of the State Council

LEGAL SYSTEM

Source: Chinese Academy of Social Sciences, 1989

LOCAL RULES AND REGULATIONS

enacted and revised by the People's Congress and its Standing Committee of Provinces, Autonomous Regions or Municipalities directly under the central government

RULES AND REGULATIONS

made by the People's Government of Provinces, Autonomous Regions and Municipalities directly under the central government

> "USE POWER FOR THE PEOPLE, SHOW COMPASSION FOR THE PEOPLE AND SEEK BENEFIT FOR THE PEOPLE."
> Hu Jintao

The opening up of China, and the development of a market-oriented economy, has been accompanied by a mushrooming of economic, social and cultural organizations promoting and protecting the interests of their members. Observers in China and elsewhere detect an incipient civil society and assert that this is associated with the development of democracy. This may well be the case but it is not inevitable.

The Party-state insists upon its control by requiring organizations to register. It can go further and impose outright bans on independent trade union movements. Equally important are the limits of tolerance as evidenced by the over-reaction to the activities of the Falun Gong. Yet the number of protests continues to rise, and the Party-state will have to seek an accommodation between freedom of expression and political stability.

DEATH PENALTY

The number of executions carried out annually in China is unknown. In 2002 Amnesty recorded 1,060, but believes the actual figure to be closer to 15,000 a year (representing 1 person in 86,000). In March 2004, a senior delegate from Chongqing Municipality estimated the figure at nearly 10,000. In October 2004 it was announced that in future the Supreme People's Court would review all death sentences, which it is hoped will reduce the number of executions.

The death penalty can be given for 68 offences, two-thirds of which are non-violent crimes such as bigamy, internet-hacking and cyber crimes, stealing petrol and tax evasion.

A party of schoolchildren aged 6 – 17 years joined an audience of 2,500 in Hunan province to witness the sentencing and execution of six men in October 2004.

INTELLECTUAL FREEDOM
1957–2004

- action encouraging freedom
- ● action discouraging freedom

Despite signing the International Covenant on Civil and Political Rights (ICCPR) in 1998, China had made no effort to ratify it by mid-2004.

1957 One Hundred Flowers
Invited by Mao Zedong to criticize the Communist Party, Chinese intellectuals attack its right to govern.

● 1958 Great Leap Forward
Crackdown on intellectuals.

● 1966–76 Cultural Revolution
Further suppression of intellectuals. Many are punished by exile to the countryside.

1978 Democracy Wall
Deng Xiaoping encourages citizens to paste political tracts on a Beijing wall.

● 1979 Wall closed
Wei Jingsheng accuses Deng Xiaoping of tyranny and is arrested. Wall closed.

● 1982–83 Spiritual pollution
Deng cracks down on greater personal freedoms demanded by *People's Daily* editor and others.

1986 Bourgeois Liberalism
Deng's renewed interest in administrative reform encourages students to demand more freedom from Communist Party control.

● 1987 Backlash
Reformist party chief Hu Yaobang sacked.

● 1989 Tiananmen Massacre
Politburo split.
Deng orders troops to attack students demonstrating for more freedom. Party chief Zhao Ziyang sacked.

● 1994 Right and Left
Neoconservatives and Party veterans call for tighter central control of economy and provinces.

1997 Sexual rights or liberation
The Chinese homosexual movement begins a step-by-step development with the establishment of a pager hotline.

1998 New Liberalism
Liberals attack neoconservatives and Party veterans, and call for greater political freedom and curbs on Communist Party power.

● 1999 Falun Gong
Mass movement perceived of as threat to state. Law passed banning "cults".

2000 Three Represents
Jiang Zemin opens party membership to entrepreneurs, intellectuals and scientists.

● 2002– Censorship
Internet search engines Google and Altavista received temporary total block; permanent selective blocking ongoing.

2004 Private protection
State constitution amended to protect private property and guarantee human rights. Law passed banning discrimination against people with HIV/AIDS.

2001–08 Olympics
In the build-up to the Beijing Olympics in 2008, international groups are pressing China to improve its human rights' record, to allow independent trade unions, to abide by anti-torture conventions, and to expand personal freedoms.

In 2002, tens of thousands of disaffected, laid-off, and unemployed workers in northeast China protested about non-payment of back wages and pensions, loss of benefits, and insufficient severance pay.

After being targeted by ACFTU, Wal-Mart agreed in 2004 to allow trade unions to be established in its 39 branches in China if requested by the workers.

GENDER RATIO
In trade unions
2002

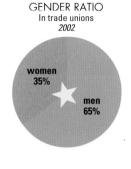

women 35%

men 65%

ACWF OFFICIALS
Full-time workers by political status
2002

total: 52,529

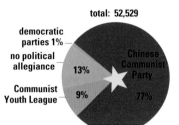

democratic parties 1%

no political allegiance 13%

Communist Youth League 9%

Chinese Communist Party 77%

The Communist Youth League of China (CYL)
Membership: 69.8 million *2002*

Led by the Communist Party of China, the CYL provides an opportunity for China's youth to study communism in practice, and implement the Party's principles.

All-China Federation of Trade Unions (ACFTU)
Membership: 133 million *2002*

Only legal trade union organization. Chinese trade unions organize both "horizontally", to form a federation of trade unions representing workers in a single enterprise, and "vertically", to form a national trade union representing a specific skill group.

All-China Women's Federation (ACWF)
Over 1 million local representatives' committees and federations *2004*

Its aim is to protect women's rights and interests and promote equality between men and women.

ORGANIZATIONS
And their relationship to the Party-state
2004

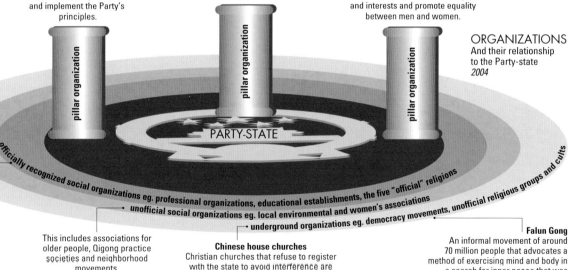

pillar organization

pillar organization

pillar organization

PARTY-STATE

officially recognized social organizations eg. professional organizations, educational establishments, the five "official" religions

unofficial social organizations eg. local environmental and women's associations

underground organizations eg. democracy movements, unofficial religious groups and cults

This includes associations for older people, Qigong practice societies and neighborhood movements.

Chinese house churches
Christian churches that refuse to register with the state to avoid interference are increasingly being treated as a cult and caught up in the backlash sparked by the rise of the Falun Gong. Members and leaders of house churches have been fined, forced to undergo "education through labor", imprisoned and threatened with the death penalty.

Falun Gong
An informal movement of around 70 million people that advocates a method of exercising mind and body in a search for inner peace that was banned by the government in 1999 as a dangerous "cult", following a demonstration in Beijing. Since then at least 1,000 people (and maybe as many as 5,000) have been tortured to death, with thousands more imprisoned in labor camps.

GONGOS
Number of Beijing-based Government Organizations and Non-Governmental Organizations by area of concern
2004

Source: Zhang, 2003

52	34	23	23	16	14	14	12	12	10	9	5	5	4
trade unions, associations for professionals, scholars, alumni	survey and research	international exchange	social services	environmental protection	poverty alleviation	vocational and adult education	sports, entertainment	legal services	community development	cultural	international aid	disaster prevention and relief	psychological counseling

Part Five

LIVING IN CHINA

CHINA'S 25-YEAR LEAP INTO MODERNIZATION has entailed great changes for the people and structure of Chinese society. The rate of transformation has been fast, and, unsurprisingly, has caused anguish as well as jubilation. Some are "getting rich quick", but others are unemployed, unsupported by welfare, and unsure of whether they count in this new entrepreneurial world. For every story of progress and reform, there is also likely to be a tale of suffering or social regression.

China's social conditions vary from one province to another, from the countryside to the towns, and the great metropolitan districts of Beijing, Shanghai, Tianjin, Shenyang, from semi-autonomous regions such as Tibet and Xinjiang, to the special administrative regions of Hong Kong and Macau. Generalizations are not sustainable in such an enormous spectrum of difference. So, how can we describe Chinese society today?

In 2004, nearly half of all mainland Chinese regarded themselves as "middle class".

China does have some clearly defined challenges. Despite the achievements of the one-child policy and sterilization programs since the 1980s, there are still nearly 1.3 billion people to be kept fed and healthy. In cities such as Beijing, professional people are finding that they and their children are in danger of joining the ranks of over-fed, under-nourished and under-exercised Western junk-food addicts. Yet on the same streets there are domestic migrants eating frugal meals whilst working long days in low-paid physical labor. These people are often rural workers, traveling without proper documentation, with no access to education for their children if they end up settling in the city, and with little or no money for health insurance, or any hope of a pension scheme. They are likely to spend years living in temporary housing, without adequate sanitation, and without access to the knowledge economy of the educated classes.

In an uncertain world religion prospers, and the opium of the people is burning steadily in China.

Christianity is growing in the south and east, whilst the Islamic faith is strong in the west. Cult religions based on meditation and charismatic leadership, of which Falun Gong is the best known, have achieved millennial status both for their followers and in the terrified eyes of the state.

Perhaps the most obvious aspect of China's transforming society is the amount and variety of media available on the streets and through television. News stands carry proliferating titles, and there is an increasing understanding of the local market. Fashion magazines range from upmarket international titles to catalog-style clothing and cosmetic listings, with lots of free gifts! Television is ubiquitous, with local stations taking a large share of the viewers' attention away from Central Chinese Television. Shanghai Oriental Pearl Company is an example of a huge media conglomerate with interests in hotels, television, animation, film and doubtless much more besides. The advertising industry is maturing, with international and local companies working on brand development and maintenance with a sure idea of the relative sophistication, and national preferences, of the market. The film industry is perhaps not as healthy as the rest of the sector. The greatest problem continues to be the explosion of televisual entertainment. People can buy cheap – often pirated – video compact discs, play them on the television at home, relax with friends, and then put in some karaoke practice. Yet, when there is a good "big screen" movie in town, then everyone goes to the brand-new multiplex.

The future of China's media is part of the story of China's social prognosis. The media are not free to print or televise whatever they want, when they want. Regulations are more stringent than in most Western societies, yet in Chinese media there is a sense of a society that is aware of its challenges, and its potential. The reach of the state is long, but the reach of social ambition is also extensive.

EATING IN THE EAST, SLEEPING IN THE WEST

China's households are changing in response to new social norms and economic demands.

While rural households are likely to have larger families, one-child families are the rule in cities. High divorce rates are breaking down the traditional family.

An urban building boom has created a vibrant housing market. One political consequence is that the urban citizen is also becoming privatized, preferring to participate in property owners' committees rather than the grass-roots community residents' committees.

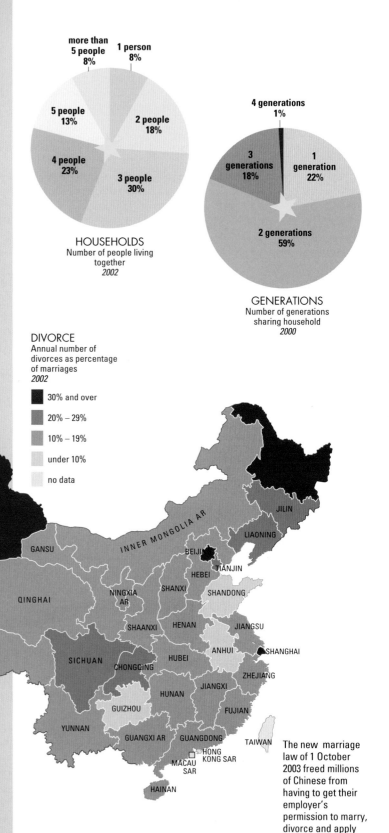

HOUSEHOLDS
Number of people living together
2002

- more than 5 people 8%
- 1 person 8%
- 2 people 18%
- 3 people 30%
- 4 people 23%
- 5 people 13%

GENERATIONS
Number of generations sharing household
2000

- 4 generations 1%
- 1 generation 22%
- 2 generations 59%
- 3 generations 18%

DIVORCE
Annual number of divorces as percentage of marriages
2002

- 30% and over
- 20% – 29%
- 10% – 19%
- under 10%
- no data

DIVORCE RATE
As percentage of total increase

- 1985 0.9%
- 1992 1.5%
- 1997 1.9%
- 2002 1.8%

The new marriage law of 1 October 2003 freed millions of Chinese from having to get their employer's permission to marry, divorce and apply for a passport.

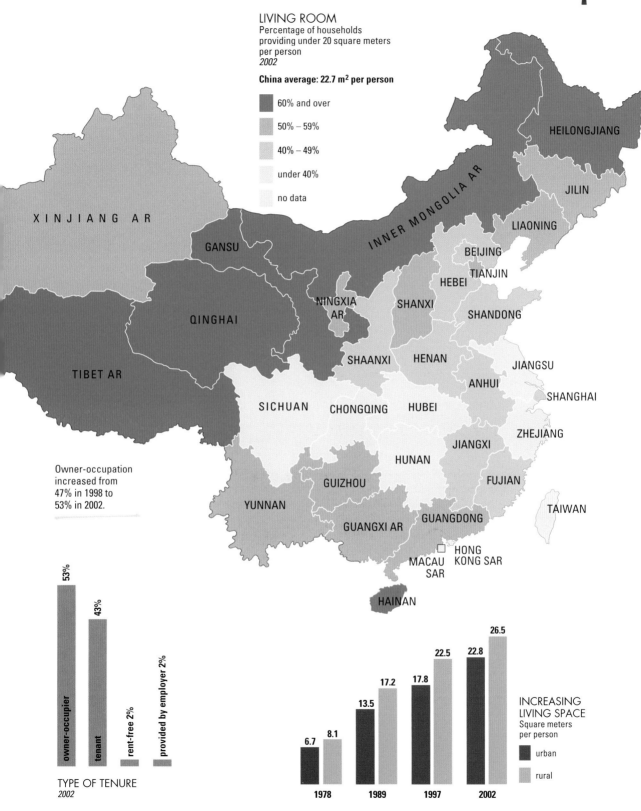

LIVING ROOM
Percentage of households
providing under 20 square meters
per person
2002

China average: 22.7 m² per person

- 60% and over
- 50% – 59%
- 40% – 49%
- under 40%
- no data

XINJIANG AR

GANSU

INNER MONGOLIA AR

HEILONGJIANG

JILIN

LIAONING

BEIJING

TIANJIN

HEBEI

NINGXIA AR

SHANXI

SHANDONG

QINGHAI

SHAANXI

HENAN

JIANGSU

TIBET AR

ANHUI

SHANGHAI

SICHUAN

CHONGQING

HUBEI

ZHEJIANG

JIANGXI

HUNAN

FUJIAN

GUIZHOU

YUNNAN

GUANGXI AR

GUANGDONG

TAIWAN

MACAU
SAR

HONG
KONG SAR

HAINAN

Owner-occupation
increased from
47% in 1998 to
53% in 2002.

53% owner-occupier

43% tenant

rent-free 2%

provided by employer 2%

TYPE OF TENURE
2002

1978: 6.7 / 8.1
1989: 13.5 / 17.2
1997: 17.8 / 22.5
2002: 22.8 / 26.5

INCREASING
LIVING SPACE
Square meters
per person

- urban
- rural

IDEOLOGY CANNOT SUPPLY RICE

People in China are consuming, on average, about 1,000 calories a day more than they did in the mid-1960s. And their diet is much more varied, with increasing amounts of meat and dairy products. Although, Chinese society retains its regional cuisine, there are creeping signs of globalization of taste and delivery, with US-branded fast-food outlets in major cities across the country.

One result of changing diets has been an increase in the number of obese and overweight people, especially among urban residents. Obesity rates in China doubled in a decade: by 2002 some 60 million people were obese and another 200 million, or 23 percent of the population, were overweight. With a rise in the number of people suffering from high blood pressure and diabetes, the government is anxious to warn people of the link between diet and health.

Coca-Cola is focusing on marketing its soft drinks to China's rural population. It aims to make China its third biggest market by 2008.

Expenditure on food in rural Tibet is 64% of living expenditure.

UNDERWEIGHT CHILDREN
Percentage of under-five-year-olds moderately or severely underweight
1995–2002

Source: UNICEF, *State of the World's Children 2004*

- 47% India
- 45% Cambodia
- 38% Bangladesh
- 33% Vietnam
- 28% Philippines
- 26% Indonesia
- 19% Thailand
- 12% Malaysia
- 11% China ★

CHANGING RURAL DIETS
Annual consumption per person of major foods in rural households
1985–2002

	1985	1997	2002
grain	466 kg	440 kg	237 kg
pork beef mutton	11 kg	13 kg	15 kg
fresh vegetables	131 kg	107 kg	111 kg
liquor	4 liters	7 liters	8 liters

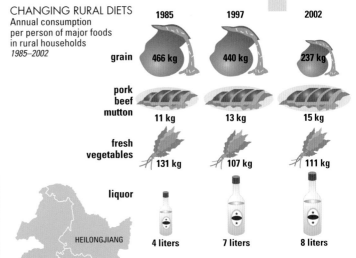

RURAL EXPENDITURE ON FOOD
As a percentage of living expenditure
2002

- 55% and over
- 50% – 54%
- 45% – 49%
- 40% – 44%
- 35% – 39%
- under 35%
- no data

XINJIANG AR
GANSU
INNER MONGOLIA AR
QINGHAI
NINGXIA AR
TIBET AR
SHANXI
SHAANXI
SICHUAN
CHONGQING
YUNNAN
GUIZHOU
GUANGXI AR
HEILONGJIANG
JILIN
LIAONING
BEIJING
TIANJIN
HEBEI
SHANDONG
HENAN
JIANGSU
ANHUI
SHANGHAI
HUBEI
ZHEJIANG
JIANGXI
HUNAN
FUJIAN
GUANGDONG
TAIWAN
MACAU SAR
HONG KONG SAR
HAINAN

URBAN EXPENDITURE ON FOOD
As a percentage of living expenditure *2002*

- 45% and over
- 40% – 44%
- 35% – 39%
- under 35%
- no data

DINING OUT
As a percentage of expenditure on food

- 20% and over
- under 15%

XINJIANG AR

GANSU

INNER MONGOLIA AR

HEILONGJIANG

JILIN

LIAONING

BEIJING

TIANJIN

NINGXIA AR

QINGHAI

HEBEI

SHANDONG

SHANXI

TIBET AR

SHAANXI

HENAN

JIANGSU

SICHUAN

CHONGQING

HUBEI

ANHUI

SHANGHAI

ZHEJIANG

HUNAN

JIANGXI

GUIZHOU

FUJIAN

YUNNAN

GUANGXI AR

GUANGDONG

TAIWAN

MACAU SAR

HONG KONG SAR

HAINAN

CHINESE CUISINE
The Four Styles

Source: authors

SHANDONG

SICHUAN

JIANGSU

GUANGDONG

China's fast-food industry is growing 20% annually. KFC opened its 600th restaurant in China in 2002.

China produces 70% of the world's farmed fish, and accounts for 30% of all fish production.

FISH PRODUCTION
1985–2002
thousand tons

	1985	1995	2002
wild seawater	3,485	10,268	14,335
farmed seawater	712	4,123	12,128
wild freshwater	476	1,373	9,408
farmed freshwater	2,378	2,252	16,930

GREAT VESSELS TAKE LONGER TO COMPLETE

The Chinese education system is arguably the most developed of any transition economy in the world. Society and parents set great store by educational opportunities and hard work. They work to pay the fees and the children work hard to finish the homework. Entry to higher education is severely competitive, even as universities flourish and expand their campuses.

Rural children do not necessarily have the dubious luxury of this pressure, however. As standards and expectations rise in wealthier parts of the country, poorer regions struggle to provide adequate resources for schools to teach even the basics to post-primary age-groups.

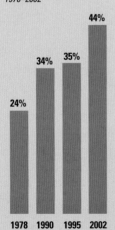

WOMEN STUDENTS
Women as percentage of students in colleges and universities
1978–2002

1978	1990	1995	2002
24%	34%	35%	44%

HIGHER EDUCATION
Breakdown of subjects being studied
2002

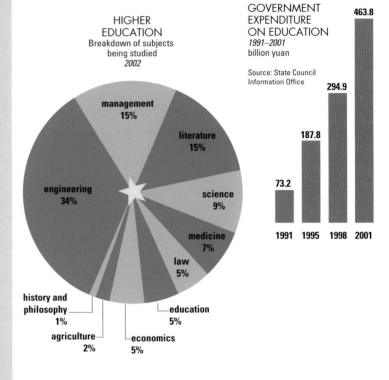

- management 15%
- literature 15%
- engineering 34%
- science 9%
- medicine 7%
- law 5%
- education 5%
- economics 5%
- agriculture 2%
- history and philosophy 1%

GOVERNMENT EXPENDITURE ON EDUCATION
1991–2001
billion yuan

Source: State Council Information Office

1991	1995	1998	2001
73.2	187.8	294.9	463.8

OVERSEAS STUDENTS
Number of students studying abroad and number returning
1992–2002

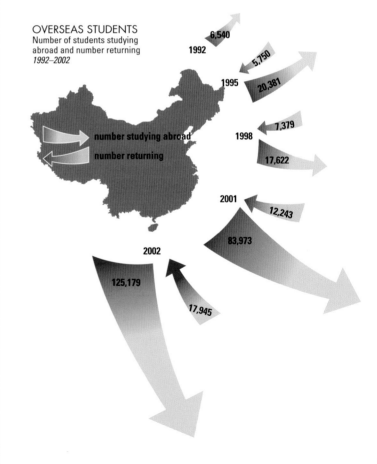

number studying abroad
number returning

Year	number studying abroad	number returning
1992	6,540	5,750
1995	20,381	7,379
1998	17,622	12,243
2001	83,973	17,945
2002	125,179	

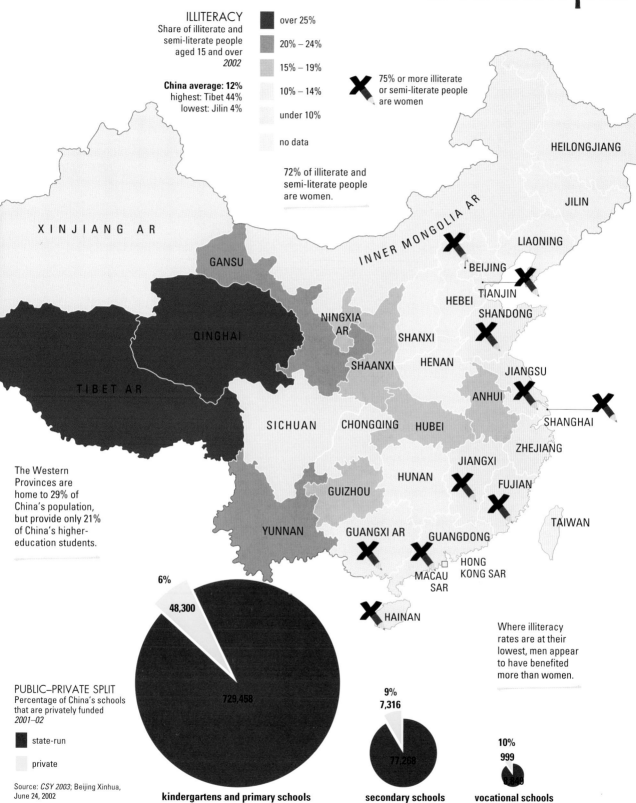

ILLITERACY
Share of illiterate and
semi-literate people
aged 15 and over
2002

China average: 12%
highest: Tibet 44%
lowest: Jilin 4%

- over 25%
- 20% – 24%
- 15% – 19%
- 10% – 14%
- under 10%
- no data

75% or more illiterate
or semi-literate people
are women

72% of illiterate and
semi-literate people
are women.

HEILONGJIANG

XINJIANG AR

JILIN

GANSU

INNER MONGOLIA AR

LIAONING

BEIJING

QINGHAI

HEBEI

TIANJIN

SHANDONG

NINGXIA AR

SHANXI

SHAANXI

HENAN

TIBET AR

JIANGSU

ANHUI

SHANGHAI

SICHUAN

CHONGQING

HUBEI

ZHEJIANG

The Western
Provinces are
home to 29% of
China's population,
but provide only 21%
of China's higher-
education students.

JIANGXI

HUNAN

FUJIAN

GUIZHOU

TAIWAN

YUNNAN

GUANGXI AR

GUANGDONG

HONG
KONG SAR

MACAU
SAR

HAINAN

Where illiteracy
rates are at their
lowest, men appear
to have benefited
more than women.

6%
48,300

729,458

PUBLIC–PRIVATE SPLIT
Percentage of China's schools
that are privately funded
2001–02

- state-run
- private

9%
7,316

77,268

10%
999

kindergartens and primary schools

secondary schools

vocational schools

Source: *CSY 2003*; Beijing Xinhua,
June 24, 2002

THE 1-2-4 PHENOMENON:
ONE CHILD, TWO PARENTS, FOUR GRANDPARENTS

By 2050 it is estimated that there will be a staggering 400 million people in China aged 65 or over, largely as a result of an increase in longevity.

The problem is that this group of people will, by then, represent 24 percent of the population. The small-family policy has ensured that the number of working-age people available to support the older age-group is declining, and the social welfare system is woefully inadequate.

PEOPLE OVER 65 YEARS OLD IN ASIA
As a percentage of total population
2002

Source: *World Development Indicators 2004*

18.2% Japan

7.8% South Korea

7.1% China ★

5.4% Vietnam
5.1% India, Indonesia

4.3% Malaysia
3.7% Philippines

The Chinese government spends 25% more on defense than on welfare and pensions. The UK government spends more than four times as much on welfare and pensions as on defense.

RURAL WELFARE RELIEF

This is a government scheme for people with no other means of support. It guarantees to address five basic needs. However, in 2002 fewer than 1.6 million people met the stringent criteria that determine who is eligible for relief.

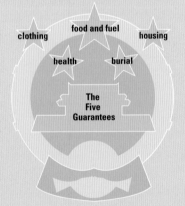

clothing
food and fuel
housing
health
burial

The Five Guarantees

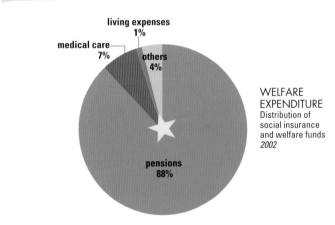

living expenses
1%

medical care
7%

others
4%

pensions
88%

WELFARE EXPENDITURE
Distribution of social insurance and welfare funds
2002

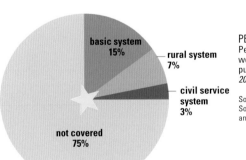

basic system
15%

rural system
7%

civil service system
3%

not covered
75%

PENSIONS
Percentage of total workforce covered by public pension system
2002

Source: Ministry of Labor and Social Security, cited by Jackson and Howe, 2004

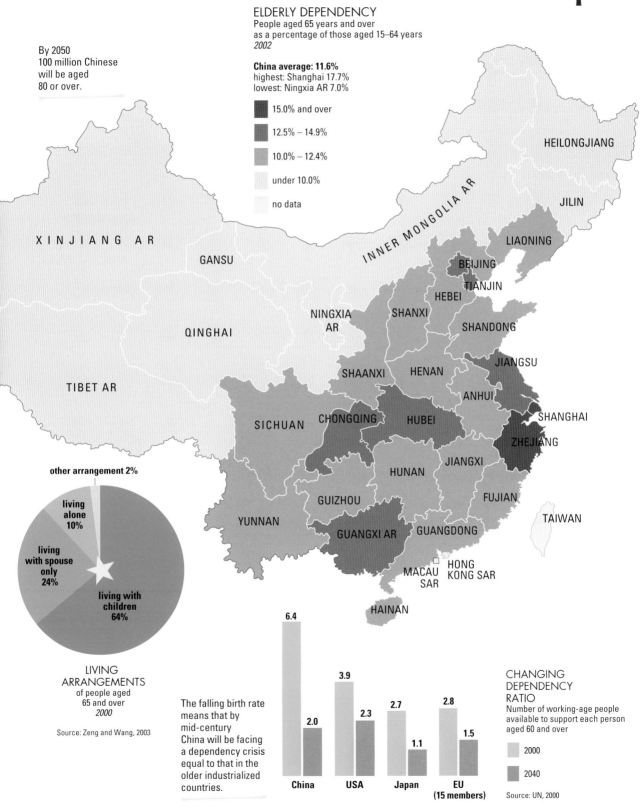

By 2050
100 million Chinese
will be aged
80 or over.

ELDERLY DEPENDENCY
People aged 65 years and over
as a percentage of those aged 15–64 years
2002

China average: 11.6%
highest: Shanghai 17.7%
lowest: Ningxia AR 7.0%

- 15.0% and over
- 12.5% – 14.9%
- 10.0% – 12.4%
- under 10.0%
- no data

HEILONGJIANG

JILIN

LIAONING

XINJIANG AR

GANSU

INNER MONGOLIA AR

BEIJING
TIANJIN

HEBEI

SHANXI

SHANDONG

NINGXIA
AR

QINGHAI

SHAANXI

HENAN

JIANGSU

TIBET AR

ANHUI

SHANGHAI

SICHUAN

CHONGQING

HUBEI

ZHEJIANG

HUNAN

JIANGXI

FUJIAN

GUIZHOU

YUNNAN

GUANGXI AR

GUANGDONG

TAIWAN

MACAU
SAR

HONG
KONG SAR

HAINAN

other arrangement 2%

living alone 10%

living with spouse only 24%

living with children 64%

LIVING ARRANGEMENTS
of people aged
65 and over
2000

Source: Zeng and Wang, 2003

The falling birth rate
means that by
mid-century
China will be facing
a dependency crisis
equal to that in the
older industrialized
countries.

6.4

3.9

2.7

2.8

2.0

2.3

1.1

1.5

China

USA

Japan

**EU
(15 members)**

CHANGING DEPENDENCY RATIO
Number of working-age people
available to support each person
aged 60 and over

- 2000
- 2040

Source: UN, 2000

SUITING THE MEDICINE TO THE ILLNESS

Is accessible healthcare a victim of China's transformation to a market-oriented economy? Prior to the economic reforms that began in 1978, China took great pride in the coverage provided by its health system. Today, healthcare is increasingly becoming a privilege of those who can afford it. In 2000, a World Health Organization survey of equality of access to medical care ranked China fourth from the bottom.

And it is not just the poor who are the victims of ill health. According to a WHO official: "There is a contradiction in China that as the country becomes wealthier it faces a whole new set of health problems related to diet, pollution, smoking and stress".

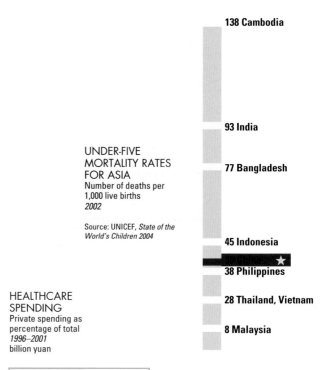

UNDER-FIVE MORTALITY RATES FOR ASIA
Number of deaths per 1,000 live births
2002

Source: UNICEF, *State of the World's Children 2004*

- 138 Cambodia
- 93 India
- 77 Bangladesh
- 45 Indonesia
- 38 China ★
- 38 Philippines
- 28 Thailand, Vietnam
- 8 Malaysia

HEALTHCARE SPENDING
Private spending as percentage of total
1996–2001
billion yuan

Year	%	billion yuan
1996	54%	285.7 billion yuan spent in total
1997	57%	338.5
1998	58%	377.7
1999	59%	417.9
2000	61%	476.4
2001	60%	515.0

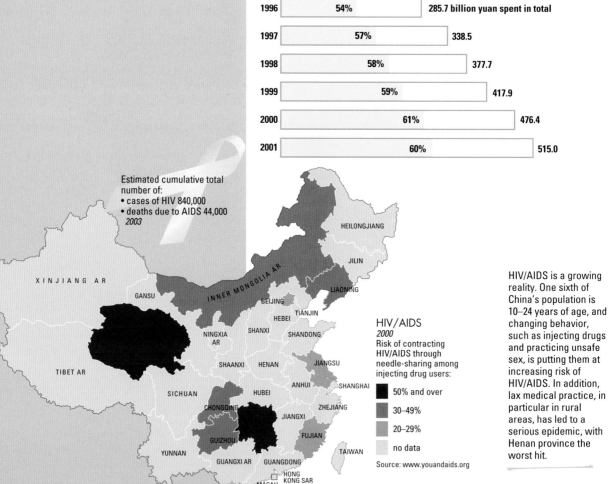

Estimated cumulative total number of:
• cases of HIV 840,000
• deaths due to AIDS 44,000
2003

HIV/AIDS
2000
Risk of contracting HIV/AIDS through needle-sharing among injecting drug users:

- 50% and over
- 30–49%
- 20–29%
- no data

Source: www.youandaids.org

HIV/AIDS is a growing reality. One sixth of China's population is 10–24 years of age, and changing behavior, such as injecting drugs and practicing unsafe sex, is putting them at increasing risk of HIV/AIDS. In addition, lax medical practice, in particular in rural areas, has led to a serious epidemic, with Henan province the worst hit.

The number of doctors, which increased steadily during the 1980s and 1990s, declined from 2.1 million in 2001 to 1.8 million in 2002.

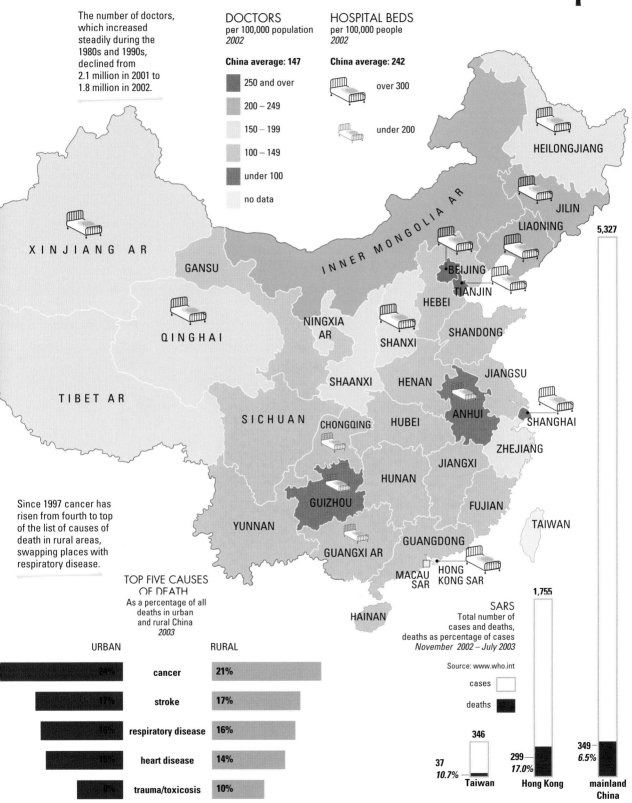

DOCTORS
per 100,000 population
2002

China average: 147

- 250 and over
- 200 – 249
- 150 – 199
- 100 – 149
- under 100
- no data

HOSPITAL BEDS
per 100,000 people
2002

China average: 242

- over 300
- under 200

Since 1997 cancer has risen from fourth to top of the list of causes of death in rural areas, swapping places with respiratory disease.

TOP FIVE CAUSES OF DEATH
As a percentage of all deaths in urban and rural China
2003

URBAN		RURAL
24%	cancer	21%
17%	stroke	17%
16%	respiratory disease	16%
15%	heart disease	14%
8%	trauma/toxicosis	10%

SARS
Total number of cases and deaths, deaths as percentage of cases
November 2002 – July 2003

Source: www.who.int

- cases
- deaths

	Taiwan	Hong Kong	mainland China
cases	346	1,755	5,327
deaths	37	299	349
%	10.7%	17.0%	6.5%

OPIUM OF THE PEOPLE

China produces 60 percent of the world's tobacco. But production is not keeping up with the ever-increasing demand for cigarettes and other tobacco products, and in 2003 China became a net importer of raw tobacco.

One in three of all cigarettes smoked in the world is smoked in China. Despite the 1995 national ban on the advertising of cigarettes on radio, television or in print, the Chinese smoke at least 1.7 trillion cigarettes every year. Although 97 percent of these are home-produced, Western tobacco companies are actively targeting China's huge and growing market. At the beginning of 2004 China was forced by the World Trade Organization to abandon its restrictions on the sale of imported tobacco products, and retailers are now able to obtain a license to sell the increasingly sought-after foreign brands. Smuggling of imported cigarettes remains rife, with black market cigarettes selling at anything up to 30 percent below the official price.

Although the health costs to the government may outweigh the revenues received, tobacco production is so integrated into the economy, and smoking so much a part of the urban male lifestyle, as to defy easy solutions. A worrying development is there are more smokers among young teenage girls than older women.

Most of the cigarettes smoked in China are manufactured by the state-owned China National Tobacco Company for home consumption.

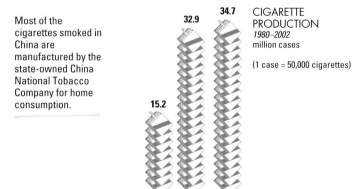

CIGARETTE PRODUCTION
1980–2002
million cases

(1 case = 50,000 cigarettes)

15.2 32.9 34.7

1980 1992 2002

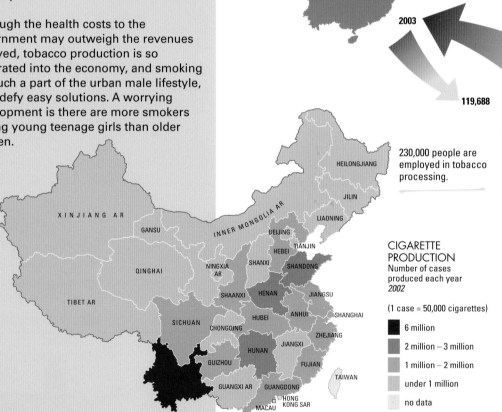

TOBACCO IMPORTS AND EXPORTS
1997–2003
metric tons

Source: USDA

imports
exports

1997   17,000   60,156

2000   56,686

113,259

2003

119,688   463,434

230,000 people are employed in tobacco processing.

CIGARETTE PRODUCTION
Number of cases produced each year
2002

(1 case = 50,000 cigarettes)

- 6 million
- 2 million – 3 million
- 1 million – 2 million
- under 1 million
- no data

HEILONGJIANG

JILIN

LIAONING

XINJIANG AR

GANSU

INNER MONGOLIA AR

BEIJING

TIANJIN

HEBEI

QINGHAI

NINGXIA AR

SHANXI

SHANDONG

SHAANXI

HENAN

JIANGSU

TIBET AR

HUBEI

ANHUI

SHANGHAI

SICHUAN

CHONGQING

ZHEJIANG

HUNAN

JIANGXI

GUIZHOU

FUJIAN

TAIWAN

GUANGXI AR

GUANGDONG

HONG KONG SAR

MACAU SAR

HAINAN

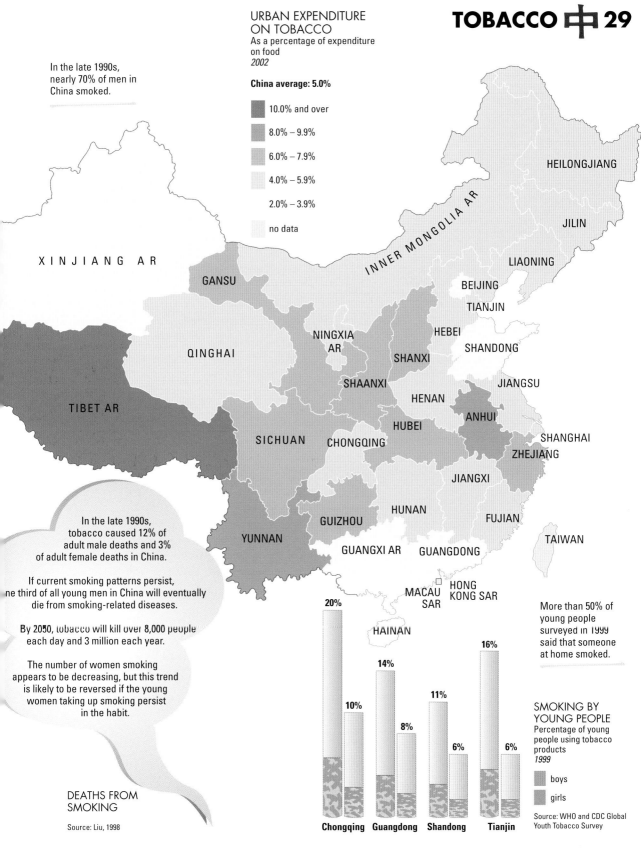

URBAN EXPENDITURE
ON TOBACCO
As a percentage of expenditure
on food
2002

China average: 5.0%

- 10.0% and over
- 8.0% – 9.9%
- 6.0% – 7.9%
- 4.0% – 5.9%
- 2.0% – 3.9%
- no data

In the late 1990s,
nearly 70% of men in
China smoked.

HEILONGJIANG

JILIN

INNER MONGOLIA AR

LIAONING

XINJIANG AR

GANSU

BEIJING

TIANJIN

NINGXIA
AR

HEBEI

SHANDONG

QINGHAI

SHANXI

SHAANXI

JIANGSU

HENAN

ANHUI

TIBET AR

HUBEI

SHANGHAI

ZHEJIANG

SICHUAN

CHONGQING

JIANGXI

In the late 1990s,
tobacco caused 12% of
adult male deaths and 3%
of adult female deaths in China.

HUNAN

FUJIAN

GUIZHOU

If current smoking patterns persist,
ne third of all young men in China will eventually
die from smoking-related diseases.

YUNNAN

GUANGXI AR

GUANGDONG

TAIWAN

By 2050, tobacco will kill over 8,000 people
each day and 3 million each year.

MACAU
SAR

HONG
KONG SAR

The number of women smoking
appears to be decreasing, but this trend
is likely to be reversed if the young
women taking up smoking persist
in the habit.

HAINAN

More than 50% of
young people
surveyed in 1999
said that someone
at home smoked.

20%

10%

14%

8%

11%

6%

16%

6%

SMOKING BY
YOUNG PEOPLE
Percentage of young
people using tobacco
products
1999

- boys
- girls

Chongqing **Guangdong** **Shandong** **Tianjin**

DEATHS FROM
SMOKING

Source: Liu, 1998

Source: WHO and CDC Global
Youth Tobacco Survey

ALTHOUGH WE LIVE SEPARATELY IN REMOTE CORNERS OF THE WORLD, WE FEEL LIKE FRIENDLY NEIGHBORS

Twenty-eight percent of children and over 83 percent of adults in Hong Kong had a cell phone in 2004. The rest of China is not far behind. The leapfrog effect of mobile technology in the Chinese market – where land lines have been too slow to compete – sees the number increasing at a rate of 4,000 new subscribers a month.

The leapfrog effect may also outpace laptop and desktop computing as 3G technology tries out the Chinese market.

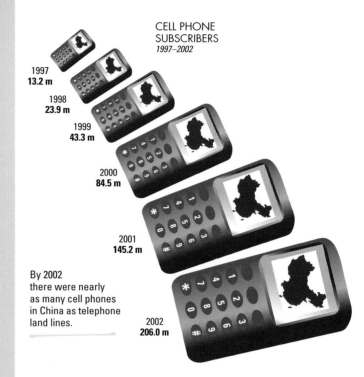

CELL PHONE SUBSCRIBERS
1997–2002

1997
13.2 m

1998
23.9 m

1999
43.3 m

2000
84.5 m

2001
145.2 m

By 2002 there were nearly as many cell phones in China as telephone land lines.

2002
206.0 m

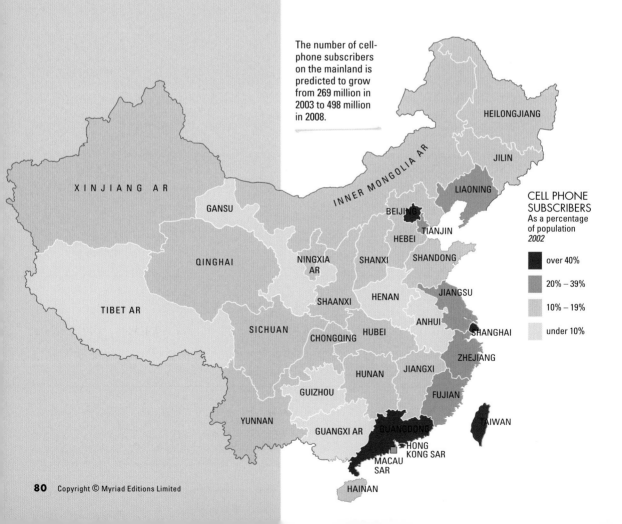

The number of cell-phone subscribers on the mainland is predicted to grow from 269 million in 2003 to 498 million in 2008.

HEILONGJIANG

JILIN

XINJIANG AR

GANSU

INNER MONGOLIA AR

LIAONING

BEIJING

TIANJIN

HEBEI

QINGHAI

NINGXIA AR

SHANXI

SHANDONG

SHAANXI

HENAN

JIANGSU

TIBET AR

ANHUI

SHANGHAI

SICHUAN

CHONGQING

HUBEI

ZHEJIANG

HUNAN

JIANGXI

GUIZHOU

FUJIAN

YUNNAN

GUANGXI AR

GUANGDONG

TAIWAN

HONG KONG SAR

MACAU SAR

HAINAN

CELL PHONE SUBSCRIBERS
As a percentage of population
2002

- over 40%
- 20% – 39%
- 10% – 19%
- under 10%

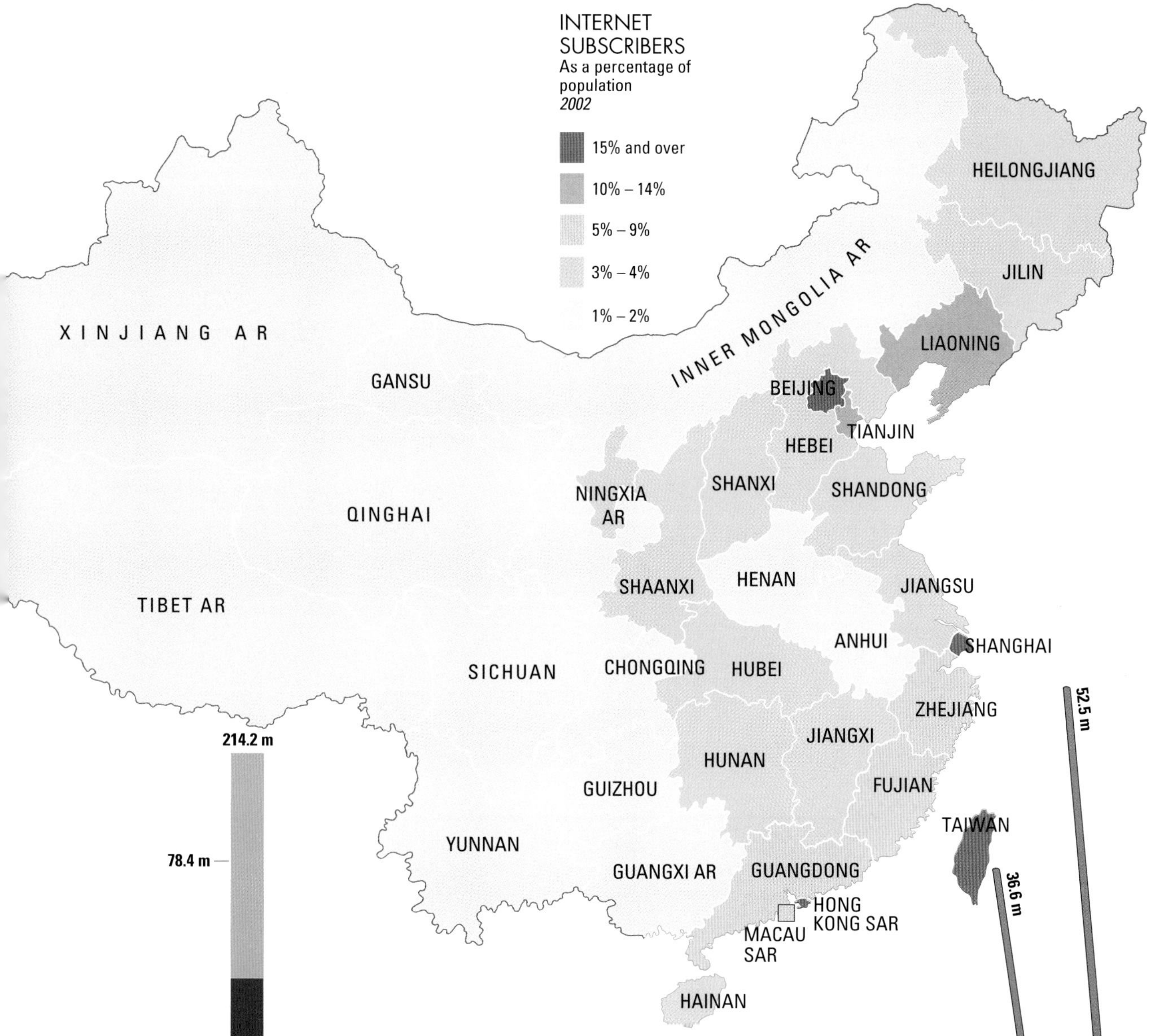

INTERNET SUBSCRIBERS

As a percentage of population
2002

- 15% and over
- 10% − 14%
- 5% − 9%
- 3% − 4%
- 1% − 2%

HEILONGJIANG

JILIN

LIAONING

XINJIANG AR

GANSU

INNER MONGOLIA AR

BEIJING

TIANJIN

HEBEI

NINGXIA AR

SHANXI

SHANDONG

QINGHAI

SHAANXI

HENAN

JIANGSU

TIBET AR

ANHUI

SHANGHAI

SICHUAN

CHONGQING

HUBEI

ZHEJIANG

JIANGXI

HUNAN

FUJIAN

GUIZHOU

TAIWAN

YUNNAN

GUANGXI AR

GUANGDONG

HONG KONG SAR

MACAU SAR

HAINAN

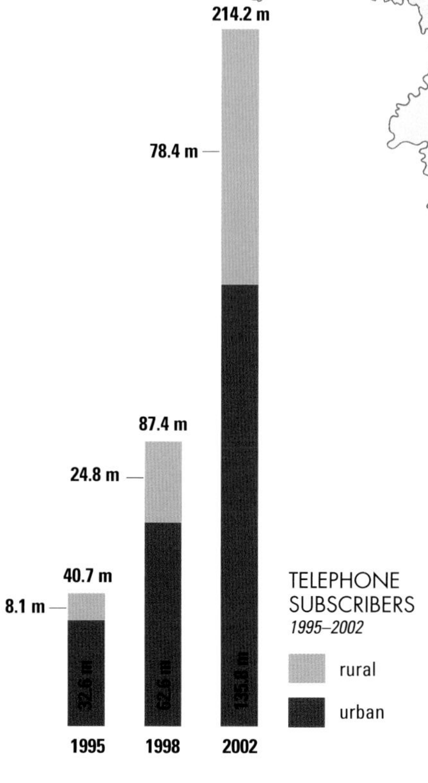

214.2 m

78.4 m

52.5 m

87.4 m

36.6 m

24.8 m

40.7 m

8.1 m

In 2004 Google launched a Chinese-language news service, but when accessed from within China it fails to list government-banned websites.

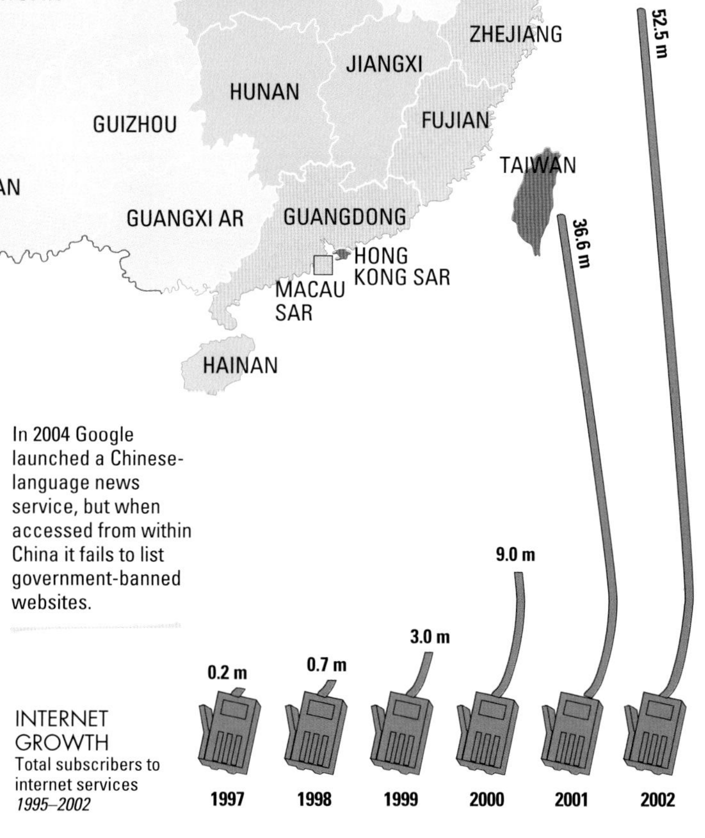

TELEPHONE SUBSCRIBERS
1995–2002

- rural
- urban

9.0 m

3.0 m

0.2 m   0.7 m

INTERNET GROWTH
Total subscribers to internet services
1995–2002

1997   1998   1999   2000   2001   2002

1995   1998   2002

THE CROWD THAT RECEIVES

The sign of a growing economy is a burgeoning media sector. Advertising on television, online, in print media and on public screens and hoardings heralds the new "creative" economy. But the true creativity of China's media industries still lies in international deals for content provision, in cross-financing conglomeration, and in working through and around the prevailing routine of censorship.

In the midst of all this, the film industry is under threat from shrinking cinema attendance, film pirating, and World Trade Organization regulations that allow competition from foreign films and merchandizing.

The Chinese Communist Party is committed to stamping out corruption in corporate and political life, and is relying on the press to help it to do this. The dilemma is that an assertive press can also undermine Party controls.

PRINT MEDIA GROWTH
Annual publications
1992–2002

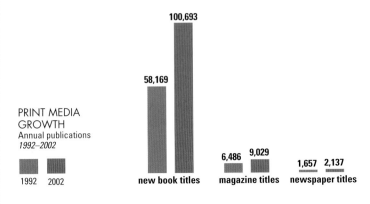

1992	2002

100,693

58,169

6,486 9,029
new book titles

1,657 2,137
magazine titles

newspaper titles

BROADCASTING
Total hours of radio and TV production
1995–2002

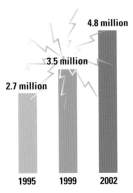

2.7 million — 1995
3.5 million — 1999
4.8 million — 2002

FILM ENTERTAINMENT
Total number of cartoons, documentary and feature films produced
1995–2002

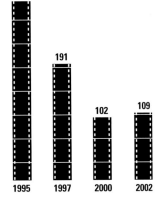

294 — 1995
191 — 1997
102 — 2000
109 — 2002

Out of 70 cinema-going nations, China ranks 66th in terms of attendance per person. The agreement made between Warner Brothers and Dalian Wanda in 2003–4 to open 30 multiplexes in shopping centers from Dalian to Nanning may turn the tide.

CINEMA AUDIENCE
Share of attendance by age group of audience
2002

Source: Zenith Media, 2002

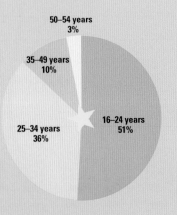

50–54 years 3%

35–49 years 10%

25–34 years 36%

16–24 years 51%

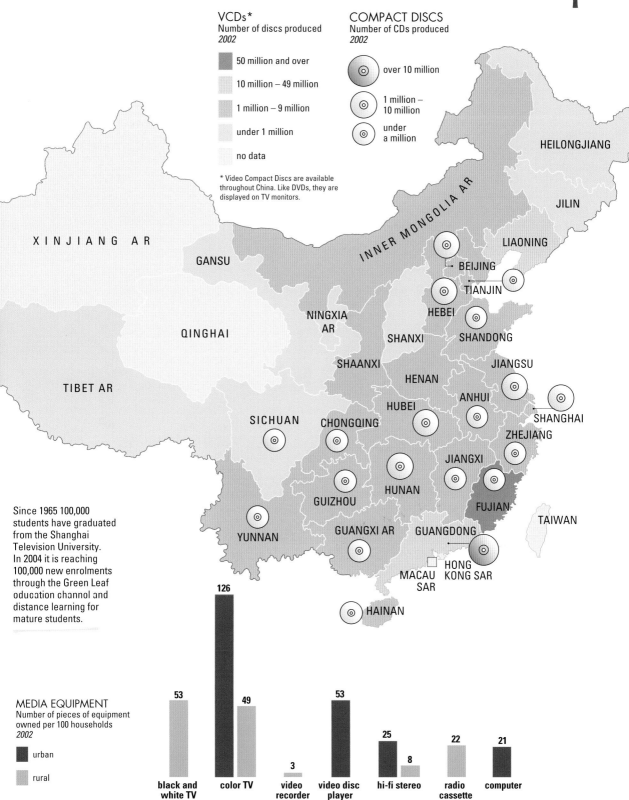

VCDs*
Number of discs produced
2002

- 50 million and over
- 10 million – 49 million
- 1 million – 9 million
- under 1 million
- no data

* Video Compact Discs are available throughout China. Like DVDs, they are displayed on TV monitors.

COMPACT DISCS
Number of CDs produced
2002

- over 10 million
- 1 million – 10 million
- under a million

HEILONGJIANG

JILIN

XINJIANG AR

INNER MONGOLIA AR

LIAONING

GANSU

BEIJING

TIANJIN

HEBEI

SHANDONG

NINGXIA AR

SHANXI

QINGHAI

SHAANXI

JIANGSU

HENAN

TIBET AR

HUBEI

ANHUI

SHANGHAI

SICHUAN

CHONGQING

ZHEJIANG

JIANGXI

HUNAN

FUJIAN

GUIZHOU

TAIWAN

YUNNAN

GUANGXI AR

GUANGDONG

MACAU SAR

HONG KONG SAR

HAINAN

Since 1965 100,000 students have graduated from the Shanghai Television University. In 2004 it is reaching 100,000 new enrolments through the Green Leaf education channel and distance learning for mature students.

MEDIA EQUIPMENT
Number of pieces of equipment owned per 100 households
2002

- urban
- rural

black and white TV	53	
color TV	126	49
video recorder	3	
video disc player	53	
hi-fi stereo	25	8
radio cassette	22	
computer	21	

MANDATES OF HEAVEN

Religions are tolerated in China under restricted conditions. Government regulations specify that no person shall use a religious venue for activities that might harm national or ethnic unity, social order, or citizens' health, or obstruct the national education system.

A 1999 law designating the Falun Gong a cult and therefore illegal could potentially be used against Christian house, or underground, churches and followers of folk religions. The crackdown on the Falun Gong resulted in the arrest of many thousands of followers.

Phuntsog Nyidron, Tibet's "Singing Nun", was released in 2004 after 15 years imprisonment but she remains under surveillance.

There are only five religions allowed by the government.

HOLIDAYS AND FESTIVALS

New Year's Day (1 day)
1 January

Chinese New Year (3–7 days)
Jan–Feb

International Women's Day
8 March

Ch'ing Ming: grave sweeping / tree planting
early April

International Labor Day (1–3 days)
1 May

Chinese Youth Day
4 May

Dragon Boat Festival
early May

International Children's Day
1 June

Founding of the People's Liberation Army
1 August

Moon Festival
Sept–Oct

National Day: Founding of the People's Republic of China (2–3 days)
1 October

RELIGIONS
Estimated number of adherents
2004

Sources: authors; press reports

- Taoist (including folk) — 250 m
- Buddhist — 100 m
- Muslim — 20 m
- Protestant Christian (registered) — 18 m
- Catholic (including underground) — 12 m

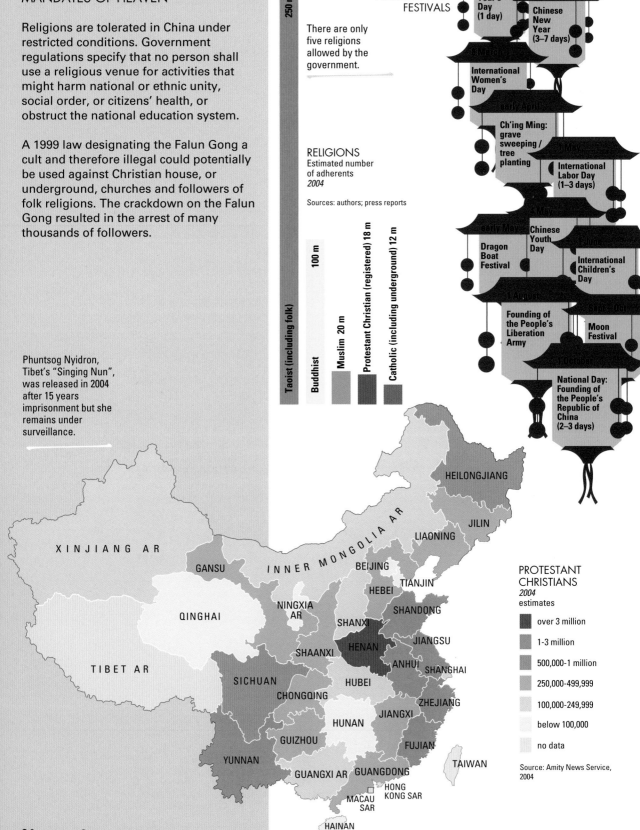

PROTESTANT CHRISTIANS
2004 estimates

- over 3 million
- 1–3 million
- 500,000–1 million
- 250,000–499,999
- 100,000–249,999
- below 100,000
- no data

Source: Amity News Service, 2004

XINJIANG AR
GANSU
INNER MONGOLIA AR
QINGHAI
NINGXIA AR
SHANXI
SHAANXI
HENAN
BEIJING
TIANJIN
HEBEI
SHANDONG
JIANGSU
ANHUI
SHANGHAI
TIBET AR
SICHUAN
CHONGQING
HUBEI
ZHEJIANG
JIANGXI
HUNAN
GUIZHOU
FUJIAN
YUNNAN
GUANGXI AR
GUANGDONG
HONG KONG SAR
MACAU SAR
HAINAN
TAIWAN
HEILONGJIANG
JILIN
LIAONING

There are 10 predominately Muslim ethnic groups, the largest of which are the Hui, estimated at nearly 10 million. The next in size are the Uygurs at 8.4 million. Kazakhs number over 1 million; the other seven ethnic groups are in the thousands.

SACRED SITES
Sites popular with tourists

🏔 major holy mountains

⭐ Communist Party sacred sites

Mao temple established 1993

Sources: O'Brien and Palmer, *The State of Religion Atlas*, 1993; press reports

The Dalai Lama, leader of Tibetan Buddhism, operates a government in exile in Dharamsala, India. The 11th incarnation of the Panchen Lama, the second highest figure in Tibet, is being held by the Chinese government, which has designated its own Panchen Lama.

HEILONGJIANG

JILIN

LIAONING

INNER MONGOLIA AR

Heng-shan
Taoist

BEIJING

TIANJIN
HEBEI

**Beijing:
Tiananmen Square.
People's Republic of China
declared 1 October 1949**

Yellow River

Gushui:
Mao's temple
joins those of
other folk gods

SHANXI

SHANDONG

XINJIANG AR

GANSU

QINGHAI

TIBET AR

NINGXIA
AR

Wu-tai-shan
Buddhist

Yan'an:
Headquarters of liberated
areas 1936-49

Tai-shan
Taoist

JIANGSU

**Shanghai:
National Congress
of Communist Party
of China 1921**

SHAANXI

HENAN

ANHUI

SHANGHAI

Hua-shan
Taoist

Song-shan
Taoist

Jiuhua-shan
Buddhist

Pu-to
Island

Pu-to-shan
Buddhist

SICHUAN

CHONGQING

HUBEI

Emei-shan
Buddhist

The Himalaya
Buddhist and Hindu

Yangtse River

GUIZHOU

Zunyi Conference.
Mao takes
over leadership 1935

HUNAN

Heng-shan
Taoist

JIANGXI

ZHEJIANG

FUJIAN

Jinggang Mountains:
Jiangxi Soviet
established 1927

YUNNAN

GUANGXI AR

GUANGDONG

HONG
KONG SAR

Guangzhou:
Peasant Movement
Institute 1926

MACAU SAP

TAIWAN

HAINAN

RUSSIA

Tuva Buryat

MONGOLIA

Inner Mongolia

NORTH
KOREA

JAPAN

Ladakh

SOUTH
KOREA

CHINA

PAKISTAN

Tibet

NEPAL BHUTAN

TAIWAN

INDIA

BANGLADESH

MACAU
SAR

HONG KONG SAR

PACIFIC
OCEAN

Maharashtra

BURMA

LAOS

THAILAND

PHILIPPINES

SRI LANKA

CAMBODIA

VIETNAM

BRUNEI

INDIAN
OCEAN

MALAYSIA

BUDDHISM
Location of the three main branches

Tibetan Buddhism

Chinese/Japanese Buddhism

Theravada Buddhism

Part Six

THE ENVIRONMENT

THE STORY OF CHINESE ECONOMIC DEVELOPMENT is a narrative of extraordinary success. It is also a narrative complicated by the dangers of rapid modernization, and by the ways in which these perils impact on the ongoing realities of Chinese everyday life. Ideally, a growing economy supports a more efficient infrastructure, smarter and cleaner industries, and the integrated management of key resources, such as water and fuel. In practice, although these ideals are certainly important to the concerned government ministries, the rapidity of change, the size of the population, and the low level of – in particular – environmental infrastructure makes the situation on the ground less rosy.

Worldwide, water is the cause of international dispute and local aggression. Access to clean drinking water is a measure of the quality and indeed possibility of people's lives. It defines borders and cultures, communities and working groups. In China, water is an increasingly scarce resource, particularly in the north and most especially in Beijing. The "north–drought, south–floods" pattern of extreme weather is predicted to continue for at least another decade. If so, it threatens the stability of both regions. The floods of the past five years have devastated villages and small towns in provinces to the far west and to the east, from Sichuan to Jiangsu. In 2003 floods around Henan City destroyed crops that should have supported 1.4 million people for the rest of the year. In 2004 the floods returned, confirming fears that the mammoth and highly controversial Three Gorges dam is not helping to control the nation's water, but instead is contributing to the chaos.

Despite these current signs of environmental imbalance, there are emerging threats to China that are likewise embedded in government thinking. The Go West campaign is a regional development strategy designed to open up the western provinces to economic growth. It is a key plank of the central government's policy for growth and for the long-term political management of ethnic minorities. There are therefore several issues at stake in the campaign, ranging from ethnic identity and local taxation to large questions concerning the hegemony of the Chinese state. But the sheer weight of development is also causing concern to commentators in the Chinese environmental agencies. The western reaches are likely to be logged and eventually deforested, water will be diverted to industrial use, and heavy industry is not adequately regulated to prevent large-scale pollution of the air, underground water sources and of the land itself. This scenario bodes ill for the severe weather pattern that has gripped China, and which is causing suffering to millions every year.

Over 10 million hectares of soil have been polluted by sewage, chemicals and dumping of industrial and domestic garbage.

There is room for hope, however. Many Chinese intellectuals, officials, farmers, and activists are aware of environmental issues, and are developing projects to improve biodiversity, halt desertification, and manage air pollution. NGOs are springing up to create educational opportunities in schools. The organization Building Hand in Hand Earth Villages engages young children in thinking about the environment through recycling for profit, which is disbursed to poorer regions. The well-connected group Friends of Nature espouses sustainable farming, and is dedicated to the protection of the wetlands in Jiangsu. These are just two of many such groups that work, in some cases very closely, with government agencies to clean up China. But the demands and stresses of individual ministries for local development and large-scale national projects are not easily overcome. As one commentator observed, Chinese agencies need to work together to solve China's problems, and to stop making new ones along the way.

THE SKY IS BLUE, THE WATER IS CLEAR, AND BEIJING IS BECOMING MORE AND MORE BEAUTIFUL

One of China's foremost dilemmas is the conflict between environmental protection and short-term economic growth. Air pollution has increased in line with industrial expansion, and with the massive rise in private car ownership. In the late 1990s, six of China's cities were judged by the World Bank to be among the ten most polluted in the world, and a clean-up campaign was launched. Some initial progress was made, but in 2003 there was a 12 percent increase in pollution compared with the previous year.

China is, however, ahead of other industrializing countries in facing up to its environmental responsibilities. The State Environmental Protection Administration (SEPA) publishes a daily city air-quality index, and an annual "dirty city" list which, in 2003, included 113 cities. A number of cities, all in coastal areas, are commended by the government for their effective pollution controls.

79% of China's infamous city smog is caused by the sulfur produced by vehicles burning low-grade petrol. Possible solutions include refitting the city's buses to run on natural gas.

CARBON DIOXIDE
EMISSIONS
per person
2000
metric tons

Source: *World Development Indicators 2004*

19.8 USA

18.0 Australia

Global warming could lead to the loss of China's extensive mountain ice-sheets, on which much of its water supply depends.

9.6 UK
9.3 Japan
9.1 South Korea

6.2 France, Malaysia

3.3 Thailand

2.2 China ★

1.3 Indonesia
1.0 Philippines
0.7 Vietnam

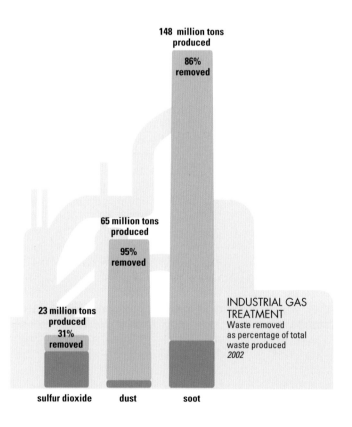

148 million tons produced
86% removed

65 million tons produced
95% removed

23 million tons produced
31% removed

INDUSTRIAL GAS
TREATMENT
Waste removed as percentage of total waste produced
2002

sulfur dioxide   dust   soot

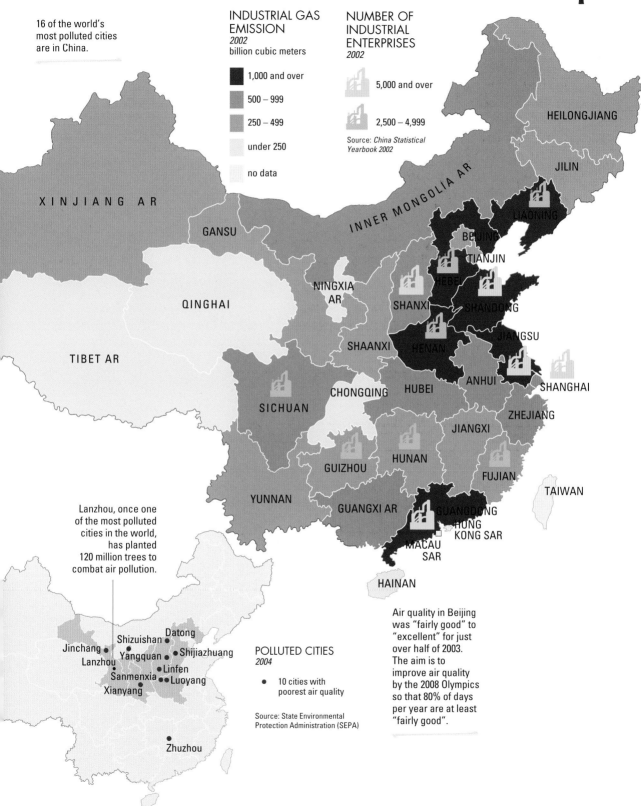

16 of the world's most polluted cities are in China.

INDUSTRIAL GAS EMISSION
2002
billion cubic meters

- 1,000 and over
- 500 – 999
- 250 – 499
- under 250
- no data

NUMBER OF INDUSTRIAL ENTERPRISES
2002

- 5,000 and over
- 2,500 – 4,999

Source: *China Statistical Yearbook 2002*

XINJIANG AR

GANSU

INNER MONGOLIA AR

HEILONGJIANG

JILIN

LIAONING

BEIJING

TIANJIN

HEBEI

QINGHAI

NINGXIA AR

SHANXI

SHANDONG

TIBET AR

SHAANXI

HENAN

JIANGSU

ANHUI

SHANGHAI

SICHUAN

CHONGQING

HUBEI

ZHEJIANG

JIANGXI

GUIZHOU

HUNAN

FUJIAN

TAIWAN

YUNNAN

GUANGXI AR

GUANGDONG

HONG KONG SAR

MACAU SAR

HAINAN

Lanzhou, once one of the most polluted cities in the world, has planted 120 million trees to combat air pollution.

Datong

Shizuishan

Jinchang

Yangquan

Shijiazhuang

Lanzhou

Linfen

Sanmenxia

Luoyang

Xianyang

Zhuzhou

POLLUTED CITIES
2004

- 10 cities with poorest air quality

Source: State Environmental Protection Administration (SEPA)

Air quality in Beijing was "fairly good" to "excellent" for just over half of 2003. The aim is to improve air quality by the 2008 Olympics so that 80% of days per year are at least "fairly good".

NOT SINKING A WELL
UNTIL ONE IS THIRSTY

Water is essential for economic and social development. While south China has water in relative abundance, the northeast is becoming increasingly water-stressed, and is relying heavily on underground aquifers, which are being used at an unsustainable rate.

China's water is controlled by over 20,000 dams – from micro-dams, which provide hydroelectricity for remote rural settlements, to the massive Three Gorges project. Yet even these feats of engineering are insufficient to tame flood waters, which appear to be striking with increasing ferocity.

Widespread droughts in 2003 affected 60 million hectares and led to crop failure on 7% of arable land.

FLOODS AND DROUGHTS
Provinces affected in 2004

- by floods
- by droughts
- unaffected

Source: www.reliefweb.int

Storms and flooding claimed over 1,300 lives and affected more than 170 million people in China between May and September 2004. Sichuan experienced the heaviest storms in 200 years.

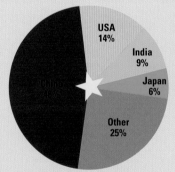

DAMS
Countries with largest shares of world's dams
2000

USA 14%
India 9%
Japan 6%
Other 25%
China

Source: The World Commission on Dams

World total: 47,655 dams

THREE GORGES DAM

height of dam:	181 meters
width of dam:	1.5 km
length of reservoir:	640 km
total capacity:	39.3 billion m³
electricity generated:	18,200 megawatts
evacuation has affected:	1.9 million people, 100 towns and villages on 44,000 hectares of land

Sources: www.chinaonline.com; www.irn.org

CHINA

HUBEI

Yunyang
Fengjie
Wushan
Badong
Wanxian
Zhong Xian
Zigui
Yangtze River
Fengdu
new reservoir 640 km (400 miles) long
Three Gorges Dam
Sandouping
Fuling
Gezhou Dam
Yichang

CHONGQING

large towns and cities affected

The northeast relies heavily on water flowing into the region, and on underground aquifers.
In Beijing, Tianjin, Hebei and Shandong the water table is dropping by 3 meters a year.

WATER AVAILABILITY
Annual renewable resource 2002 cubic kilometers

- 200 and over
- 100 – 199
- 50 – 99
- 10 – 49
- under 10
- no data

WATER USED
As a percentage of annual renewable resource 2002

- 100% and over
- 50% – 99%

WATER FOR COAL
A shortage of water in Shanxi province has cost China US$600 million a year in lost coal production, and restricted the citizens of Taiyuan to only 25 liters of water a day. Diversion of water from the Yellow River to Taiyuan through 300 km of tunnels to an elevation of 636 meters has helped alleviate the problem, but will exacerbate shortages already experienced downstream.

Work is underway on a grand scheme to supplement the Yellow River with water from the Yangtze. The eastern route is projected to be finished in 2007, the middle route, which includes 1,200 km of canals, in 2030, and the high-altitude western route in 2050.

WATER FOR THE NORTH
- river
- provincial boundary
- South-to-North water diversion:
- eastern route
- middle route
- western routes

Source: www.water-technology.net

WHEN YOU EAT FRUIT, THINK OF THE TREE THAT BORE IT. WHEN YOU DRINK WATER, THINK OF ITS SOURCE.

An increasing number of urban residents have access to tap water, but at least 20 percent in central China have to rely on less convenient, and less healthy, sources of water. Rural residents fare much worse and – probably as a consequence – use markedly less water in their homes than do their urban neighbors.

Provision for adequate sanitation arrangements is by no means universal. In 2001 it was estimated that only 69 percent of urban residents and 27 percent of rural residents had access to such facilities.

Pollution by industrial and domestic wastewater diminishes the drinking water available, and contributes to China's growing water-shortage crisis.

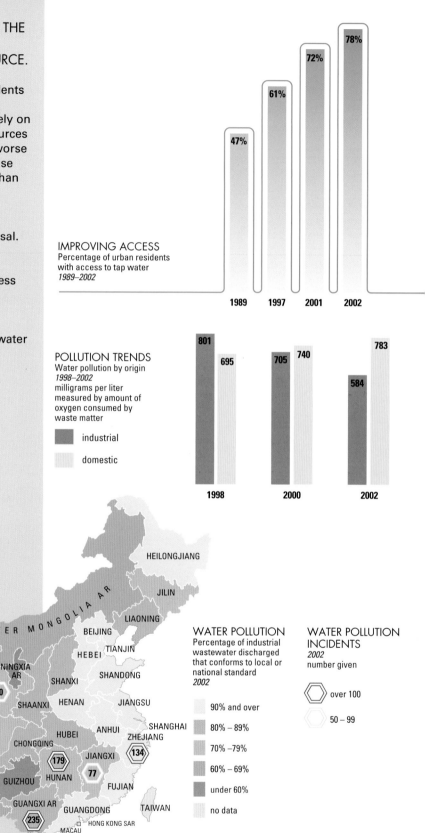

IMPROVING ACCESS
Percentage of urban residents with access to tap water
1989–2002

47%	61%	72%	78%
1989	1997	2001	2002

POLLUTION TRENDS
Water pollution by origin
1998–2002
milligrams per liter measured by amount of oxygen consumed by waste matter

- industrial
- domestic

1998	2000	2002
801 / 695	705 / 740	584 / 783

WATER POLLUTION
Percentage of industrial wastewater discharged that conforms to local or national standard
2002

- 90% and over
- 80% – 89%
- 70% –79%
- 60% – 69%
- under 60%
- no data

WATER POLLUTION INCIDENTS
2002
number given

- over 100
- 50 – 99

Map labels: HEILONGJIANG, JILIN, LIAONING, INNER MONGOLIA AR, XINJIANG AR, GANSU, BEIJING, TIANJIN, HEBEI, NINGXIA AR, SHANXI, SHANDONG, QINGHAI, SHAANXI, HENAN, JIANGSU, TIBET AR, SICHUAN, HUBEI, ANHUI, SHANGHAI, ZHEJIANG, CHONGQING, JIANGXI, HUNAN, GUIZHOU, FUJIAN, YUNNAN, GUANGXI AR, GUANGDONG, TAIWAN, HONG KONG SAR, MACAU SAR, HAINAN

Map incident values: 50, 64, 179, 77, 134, 235

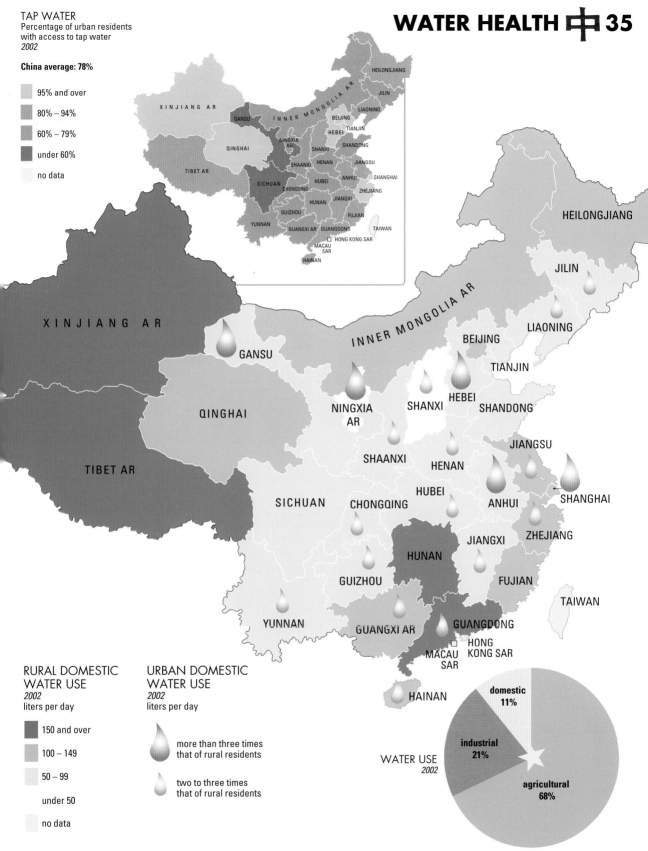

TAP WATER
Percentage of urban residents
with access to tap water
2002

China average: 78%

- 95% and over
- 80% – 94%
- 60% – 79%
- under 60%
- no data

HEILONGJIANG

XINJIANG AR

GANSU

INNER MONGOLIA AR

QINGHAI

TIBET AR

NINGXIA
AR

SHANXI

HEBEI

BEIJING

TIANJIN

SHANDONG

JILIN

LIAONING

SHAANXI

HENAN

JIANGSU

SICHUAN

CHONGQING

HUBEI

ANHUI

SHANGHAI

ZHEJIANG

HUNAN

JIANGXI

FUJIAN

GUIZHOU

YUNNAN

GUANGXI AR

GUANGDONG

HONG
KONG SAR

MACAU
SAR

HAINAN

TAIWAN

RURAL DOMESTIC
WATER USE
2002
liters per day

- 150 and over
- 100 – 149
- 50 – 99
- under 50
- no data

URBAN DOMESTIC
WATER USE
2002
liters per day

- more than three times
  that of rural residents
- two to three times
  that of rural residents

WATER USE
2002

domestic
11%

industrial
21%

agricultural
68%

需要您的配合与支持！
We Need Your Cooperation&Support!

中国经济普查
CHINA ECONOMIC CENSUS

Part Seven
TABLES

No country in the world is more aware than China of the need for good economic and social data. Such data serves as a basis for policy-making, as a set of indicators for a market-led economy, and allows commentators, both Chinese and foreign, to assess the economic and social conditions of contemporary China. How far China has managed to gather such data, and how the data has been used, needs careful consideration.

The Great Leap Forward (1957 to 1961) was an unprecedented historical lesson. A series of programs was launched, including the mass mobilization of the rural population to build infrastructure projects, and the rapid collectivization of the countryside into communes. But these ambitious plans went awry: Beijing lost control, having devolved the collection of statistics and authority for planning to the provinces and to regional levels of government. Ever-increasing targets were set, and reports of over-achievements were rife. The ensuing statistical crisis, along with the withdrawal of Soviet aid and bad weather, precipitated a famine in which up to 30 million people died. It was an experience not to be repeated.

"Seek truth from facts."
Mao Zedong

The reformers, led by Deng Xiaoping, undertook a massive gamble in 1978 by dismantling much of the command economy and replacing it with a market-led economy. At the same time, it abandoned the Maoist policy of self-sufficiency and opened up China to foreign investment. But this still required planning, and planning needs reliable data that also meets the expectations of those trading and investing in China. As the modernization of China proceeds apace and the country strives to become a powerhouse driven by trade in the international community, so must its information base improve.

This book has used two main sources of official data. The *China Statistical Yearbook* contains a wealth of information about almost every aspect of China, except the military: employment, agriculture, trade, schooling, healthcare, housing. More detailed, and more intimate, data, are given in the *China Population Statistics Yearbook*, including the number of rooms in a dwelling, and the main form of contraception used by householders. Much of this is based on the 2000 Census, the fifth in a 10-yearly series.

Taking a census of China's almost 1.3 billion people is no mean feat. The 2000 Census used 10,000 tons of paper for the questionnaires, and 5 million enumerators to collect the data. It was most notable, however, for the innovations introduced since the previous census. For the first time the questionnaire included a confidentiality clause, presumably intended to encourage respondents to be more truthful. Unusually detailed questions were asked about housing – the age and condition of the property, its water supply and sanitary facilities, the length of tenure and the weekly rent – information clearly of interest to a government working through the transition from a state-controlled to a market-led housing policy. The issue of unemployment and welfare support was addressed openly for the first time. And the timing of the census was shifted from 1 July to 1 November, with the result that migrant workers were more likely to have been enumerated at their place of work, rather than in their home town or village.

Officially produced data has two uses: as a basis for policy-making and as a way of presenting China in the best light possible. The information the Party-state chooses to collect is likely to support its promotion of a market-led economy.

We have relied mainly on official data, well aware of their strengths and weaknesses. We are confident, however, of their usefulness in providing an overview, and as a guide to trends and developments in one of the most diverse and rapidly changing countries in the world.

Selected countries	1 Population			2 Life expectancy at birth	3 Gross Domestic Product		
	Total 2003 million	Annual growth rate 2003	Density 2003 people per km²	2002	Total 2002 US$ billion	Annual growth 2002	Per person International $ 2002
Argentina	38	1.2%	14	74.3	102	−10.9%	10,880
Australia	20	1.0%	3	79.2	409	2.7%	28,260
Bangladesh	144	2.1%	999	62.1	48	4.4%	1,700
Brazil	176	1.3%	21	68.6	452	1.5%	7,770
Canada	31	0.8%	3	79.2	714	3.3%	29,480
China	**1,285**	**0.6%**	**134**	**70.7**	**1,266**	**8.0%**	**4,580**
Czech Republic	10	−0.1%	130	75.0	70	2.0%	15,780
Egypt	71	2.0%	70	68.9	90	3.0%	3,810
France	60	0.5%	109	79.2	1,431	1.2%	26,920
Germany	82	0.1%	231	78.1	1,984	0.2%	27,100
India	1,050	1.6%	319	63.4	510	4.6%	2,670
Indonesia	217	1.3%	114	66.7	173	3.7%	3,230
Israel	6	1.7%	287	78.7	104	−0.8%	19,530
Italy	57	−0.1%	191	78.4	1,184	0.4%	26,430
Japan	127	0.2%	337	81.6	3,993	0.3%	26,940
Jordan	5	2.1%	55	72.0	9	4.9%	4,220
Kazakhstan	16	−0.3%	6	61.7	25	9.8%	5,870
Malaysia	24	2.0%	73	72.8	95	4.1%	9,120
Mexico	102	1.5%	52	73.6	637	0.9%	8,970
Mongolia	3	1.2%	2	65.5	1	4.0%	1,710
New Zealand	4	0.8%	14	78.4	59	4.3%	21,740
Nigeria	121	2.6%	131	45.3	44	−0.9%	860
Pakistan	150	2.5%	188	63.8	59	2.8%	1,940
Philippines	79	1.9%	262	69.8	78	4.4%	4,170
Poland	39	−0.1%	124	73.8	189	1.4%	10,560
Romania	22	−0.2%	94	70.0	46	4.3%	6,560
Russian Federation	144	−0.5%	8	65.8	347	4.3%	8,230
Singapore	4	1.9%	6,747	–	87	2.2%	24,040
South Africa	40	−0.2%	33	46.5	104	3.0%	10,070
South Korea	47	0.6%	478	73.9	477	6.3%	16,950
Thailand	62	1.0%	121	69.2	127	5.4%	7,010
Turkey	70	1.5%	91	69.9	184	7.8%	6,390
United Kingdom	59	0.3%	244	77.5	1,566	1.8%	26,150
USA	291	1.0%	30	77.3	10,383	2.4%	35,750
Vietnam	80	1.4%	242	69.7	35	7.0%	2,300
World	**6,225**	**1.3%**	**46**	**66.7**	**32,312**	**1.9%**	**7,868**

Sources: **Col 1** FAO and UN Population Division databases; **Col 2** World Bank database April 2004; **Col 3** World Bank database April 2004, and International Monetary Fund database

Agriculture 2001 or latest percent	Industry 2001 or latest percent	Services 2001 or latest percent	Imports 2003 US$ million	Exports 2003 US$ million	China's exports 2003 US$ million	China's imports 2003 US$ million	Foreign Direct Investment in China 2003 US$ million	Selected countries
0.4%	22.9%	76.3%	138	293	447	2,729	18.9	Argentina
4.9%	20.9%	74.1%	886	704	6,264	7,300	592.5	Australia
62.1%	10.3%	23.5%	97	68	1,335	33	3.1	Bangladesh
20.6%	20.0%	59.2%	507	731	2,143	5,842	16.7	Brazil
2.9%	22.7%	74.4%	2,456	2,721	5,632	4,374	563.5	Canada
50.0%	**22.3%**	**27.7%**	**4,128**	**4,382**	–	–	–	**China**
4.8%	40.4%	54.8%	513	487	1,282	297	12.5	Czech Republic
29.6%	21.3%	49.1%	133	58	937	153	3.3	Egypt
1.6%	24.4%	74.1%	3,884	3,847	7,294	6,099	604.3	France
2.6%	32.5%	64.7%	6,017	7,484	17,442	24,292	857.0	Germany
66.7%	12.9%	20.3%	697	547	3,343	4,251	15.9	India
43.8%	17.0%	37.5%	324	607	4,482	5,747	150.1	Indonesia
19.3%	23.4%	56.0%	364	316	1,141	690	15.7	Israel
5.3%	32.1%	62.5%	2,890	2,902	6,652	5,081	316.7	Italy
4.9%	30.5%	63.9%	3,830	4,719	59,409	74,148	5,054.2	Japan
–	–	–	56	30	464	61	6.2	Jordan
22.0%	18.3%	59.8%	83	129	1,572	1,720	0.7	Kazakhstan
18.4%	32.2%	49.5%	811	1,007	6,141	13,986	251.0	Malaysia
17.6%	26.0%	56.0%	1,790	1,653	3,267	1,677	5.6	Mexico
48.9%	14.1%	14.6%	8	5	156	284	0.2	Mongolia
9.1%	22.8%	67.9%	186	165	803	1,024	65.8	New Zealand
2.9%	22.0%	75.1%	109	203	1,786	72	20.8	Nigeria
48.4%	18.0%	33.5%	130	119	1,855	575	3.4	Pakistan
37.4%	15.6%	47.0%	393	371	3,093	6,307	220.0	Philippines
19.1%	30.5%	50.4%	669	523	1,620	359	3.6	Poland
42.3%	26.2%	31.5%	240	176	506	470	15.3	Romania
11.8%	29.4%	58.8%	745	1,352	6,030	9,728	54.3	Russian Federation
0.3%	25.4%	74.2%	1,279	1,441	8,864	10,485	2,058.4	Singapore
10.9%	25.1%	60.9%	381	365	2,029	1,840	32.5	South Africa
10.3%	27.4%	62.3%	1,788	1,943	20,095	43,128	4,488.5	South Korea
46.6%	19.5%	33.9%	757	803	3,828	8,827	173.5	Thailand
32.6%	24.3%	43.1%	677	466	2,065	533	12.7	Turkey
1.4%	24.9%	73.4%	3,883	3,039	10,824	3,570	742.5	United Kingdom
2.4%	22.4%	75.2%	13,056	7,240	92,467	33,866	4,198.5	USA
69.1%	12.5%	18.7%	240	197	3,183	1,457	3.3	Vietnam
37.5%	**21.8%**	**34.4%**	**77,650**	**74,820**	**438,228**	**412,760**	**53,504.7**	**World**

Provinces	1 Population *2002* *million*	2 Annual population growth rate *2002*	3 Number of girls born per 100 boys *2000*	4 Life expectancy *2000*
Beijing	14.2	0.09%	90	76.1
Tianjin	10.1	0.15%	89	74.9
Hebei	67.4	0.53%	88	72.5
Shanxi	32.9	0.67%	89	71.7
Inner Mongolia AR	23.8	0.37%	92	69.9
Liaoning	42.0	0.13%	89	73.3
Jilin	27.0	0.32%	90	73.1
Heilongjiang	38.1	0.25%	91	72.4
Shanghai	16.3	−0.05%	90	78.1
Jiangsu	73.8	0.22%	86	73.9
Zhejiang	46.5	0.38%	88	74.7
Anhui	63.4	0.60%	78	71.9
Fujian	34.7	0.58%	85	72.6
Jiangxi	42.2	0.87%	87	69.0
Shandong	90.8	0.46%	89	73.9
		0.00%		
Henan	96.1	0.60%	84	71.5
Hubei	59.9	0.22%	78	71.1
Hunan	66.3	0.49%	79	70.7
Guangdong	78.6	0.82%	77	73.3
Guangxi AR	48.2	0.70%	80	71.3
Hainan	8.0	0.95%	74	72.9
Chongqing	31.1	0.33%	87	71.7
Sichuan	86.7	0.39%	86	71.2
Guizhou	38.4	1.08%	93	66.0
Yunnan	43.3	1.06%	92	65.5
Tibet AR	2.7	1.28%	97	64.4
Shaanxi	36.7	0.41%	82	70.1
Gansu	25.9	0.67%	87	67.5
Qinghai	5.3	1.17%	91	66.0
Ningxia AR	5.7	1.16%	92	70.2
Xinjiang AR	19.1	1.09%	94	67.4
China	**1,284.5**	**0.65%**	**86**	**71.4**

Sources: **Cols 1, 2, 4** *China Statistical Yearbook 2003;* **Col 3** *China Population Statistics Yearbook 2002*

5 **Dependency** People aged 65 and over as % of those aged 15–64 *2002*	6 **Urbanization** Urban population as % of population *2000*	7 **Ethnicity** Minority nationalities as % of total population *2000*	8 **Literacy** % of people aged 15 and over who are illiterate or semi-literate *2002*	**Provinces**
13.9%	77.5%	4.3%	5.4%	Beijing
14.4%	72.0%	2.7%	6.7%	Tianjin
10.7%	26.1%	4.4%	7.8%	Hebei
10.0%	34.9%	0.3%	6.4%	Shanxi
9.7%	42.7%	20.8%	13.5%	Inner Mongolia AR
10.6%	54.2%	16.1%	5.2%	Liaoning
8.7%	49.7%	9.2%	4.4%	Jilin
8.3%	51.5%	4.9%	6.5%	Heilongjiang
17.7%	88.3%	0.6%	8.2%	Shanghai
13.9%	41.5%	0.4%	14.3%	Jiangsu
15.4%	48.7%	0.9%	13.5%	Zhejiang
12.2%	27.8%	0.7%	17.9%	Anhui
10.7%	41.6%	1.7%	13.7%	Fujian
10.8%	27.7%	0.3%	10.8%	Jiangxi
11.7%	38.0%	0.7%	11.2%	Shandong
10.7%	23.2%	1.3%	9.1%	Henan
12.9%	40.2%	4.4%	15.1%	Hubei
12.1%	29.8%	10.1%	8.4%	Hunan
11.5%	55.0%	1.5%	7.0%	Guangdong
12.5%	28.2%	38.0%	9.5%	Guangxi AR
11.4%	40.1%	17.4%	8.9%	Hainan
12.8%	33.1%	6.5%	10.3%	Chongqing
12.2%	26.7%	5.0%	13.6%	Sichuan
10.4%	23.9%	37.8%	18.7%	Guizhou
10.3%	26.7%	33.4%	23.1%	Yunnan
9.4%	23.9%	93.9%	43.8%	Tibet AR
11.5%	23.3%	0.5%	15.6%	Shaanxi
9.1%	24.0%	8.8%	21.1%	Gansu
7.9%	34.8%	46.0%	24.8%	Qinghai
7.0%	32.4%	34.6%	17.5%	Ningxia AR
8.9%	33.8%	59.4%	8.2%	Xinjiang AR
11.6%	**36.1%**	**8.4%**	**11.6%**	**China**

Sources: **Cols 5, 7, 8** *China Statistical Yearbook 2003*; **Col 6** William Lavely, University of Washington

Provinces	9 Gross Domestic Product per person 2002 yuan	10 Value of commodity imports and exports 2002 US$ million	11 Foreign Direct Investment 2002 US$ million	12 Foreign exchange earnings from tourism 2002 US$ million
Beijing	28,449	26,702	267	3,115
Tianjin	22,380	22,850	229	342
Hebei	9,115	6,829	68	167
Shanxi	6,146	3,597	36	75
Inner Mongolia AR	7,241	2,665	27	149
Liaoning	12,986	23,426	234	550
Jilin	8,334	4,074	41	86
Heilongjiang	10,184	4,688	47	297
Shanghai	40,646	72,251	723	2,275
Jiangsu	14,391	74,489	745	1,050
Zhejiang	16,838	46,355	464	928
Anhui	5,817	4,205	42	124
Fujian	13,497	30,329	303	1,100
Jiangxi	5,829	1,997	20	72
Shandong	11,645	37,370	374	472
Henan	6,436	3,730	37	145
Hubei	8,319	4,532	45	284
Hunan	6,565	3,273	33	311
Guangdong	15,030	225,451	2,255	5,091
Guangxi AR	5,099	2,607	26	321
Hainan	7,803	1,793	18	92
Chongqing	6,347	2,023	20	218
Sichuan	5,766	4,461	45	200
Guizhou	3,153	980	10	80
Yunnan	5,179	2,328	23	419
Tibet AR	6,093	125	1	52
Shaanxi	5,523	2,784	28	351
Gansu	4,493	1,038	10	54
Qinghai	6,426	234	2	10
Ningxia AR	5,804	494	5	2
Xinjiang AR	8,382	3,082	31	99
China	**8,184**	**620,766**	**52,743**	**20,385**

13 Agriculture		14 Industry		15 Services		Provinces
Value 2002 billion yuan	Output as % of GDP 2002	Value 2002 billion yuan	Output as % of GDP 2002	Value 2002 billion yuan	Output as % of GDP 2002	
9.8	3%	111.7	35%	199.8	62%	Beijing
8.4	4%	100.2	49%	96.5	47%	Tianjin
95.7	16%	304.6	50%	212.0	35%	Hebei
19.8	10%	108.4	54%	73.6	37%	Shanxi
37.5	22%	72.8	42%	63.1	36%	Inner Mongolia AR
59.0	11%	261.0	48%	225.8	41%	Liaoning
44.6	20%	97.8	44%	82.2	37%	Jilin
44.7	12%	216.9	56%	126.6	33%	Heilongjiang
8.8	2%	256.5	47%	275.6	51%	Shanghai
111.9	11%	555.1	52%	396.2	37%	Jiangsu
69.4	9%	398.2	51%	312.0	40%	Zhejiang
77.3	22%	155.2	43%	124.4	35%	Anhui
66.5	14%	216.0	46%	185.7	40%	Fujian
53.6	22%	95.2	39%	96.3	39%	Jiangxi
139.0	13%	531.0	50%	385.3	37%	Shandong
128.8	21%	295.1	48%	192.9	31%	Henan
70.7	14%	244.6	49%	182.3	37%	Hubei
84.7	20%	173.7	40%	175.6	41%	Hunan
103.3	9%	593.6	50%	480.1	41%	Guangdong
59.6	24%	86.4	35%	99.6	41%	Guangxi AR
22.9	38%	12.5	21%	25.0	41%	Hainan
31.6	16%	82.8	42%	82.8	42%	Chongqing
102.8	21%	198.2	41%	186.5	38%	Sichuan
28.1	24%	47.5	40%	43.0	36%	Guizhou
47.1	21%	95.1	43%	81.0	36%	Yunnan
4.0	25%	3.3	20%	8.9	55%	Tibet AR
30.4	15%	92.6	45%	80.6	40%	Shaanxi
21.4	18%	53.0	46%	41.7	36%	Gansu
4.5	13%	15.4	45%	14.2	42%	Qinghai
5.3	16%	15.1	46%	12.5	38%	Ningxia AR
30.5	19%	67.2	42%	62.1	39%	Xinjiang AR
1,611.7	**15%**	**5,354.1**	**51%**	**3,513.3**	**34%**	**China**

Provinces	16 Rural household consumption as % of urban 2002	17 Expenditure on food as % of living expenditure		18 Employment	
		Rural 2002	Urban 2002	Average wage 2002 yuan	Urban unemployment rate 2002
Beijing	39%	34%	34%	21,852	1.4%
Tianjin	42%	38%	36%	16,258	3.9%
Hebei	32%	39%	35%	10,032	3.6%
Shanxi	27%	44%	33%	9,357	3.4%
Inner Mongolia AR	31%	43%	32%	9,683	4.1%
Liaoning	34%	45%	39%	11,659	6.5%
Jilin	33%	44%	36%	9,990	3.6%
Heilongjiang	27%	42%	36%	9,926	4.9%
Shanghai	46%	35%	39%	23,959	4.8%
Jiangsu	40%	40%	40%	13,509	4.2%
Zhejiang	38%	41%	40%	18,785	4.2%
Anhui	33%	47%	43%	9,296	4.0%
Fuijan	53%	46%	46%	13,306	4.2%
Jiangxi	37%	50%	41%	9,262	3.4%
Shandong	38%	42%	34%	11,374	3.6%
Henan	30%	48%	34%	9,174	2.9%
Hubei	29%	50%	37%	9,611	4.3%
Hunan	33%	53%	36%	10,967	4.0%
Guangdong	28%	48%	38%	17,814	3.1%
Guangxi AR	27%	52%	41%	10,774	3.7%
Hainan	43%	59%	45%	9,480	3.1%
Chongqing	21%	56%	38%	10,960	4.1%
Sichuan	30%	54%	40%	11,183	4.5%
Guizhou	25%	58%	39%	9,810	4.1%
Yunnan	32%	56%	42%	11,987	4.0%
Tibet AR	11%	64%	41%	24,766	4.9%
Shaanxi	23%	38%	34%	10,351	3.3%
Gansu	20%	46%	35%	11,147	3.2%
Qinghai	26%	48%	37%	14,472	3.6%
Ningxia AR	26%	45%	35%	11,640	4.4%
Xinjiang AR	24%	49%	34%	11,605	3.7%
China	28%	46%	38%	12,422	4.0%

CHINA: LIVING AND LIFESTYLE 中

19 Health		20 Number of private passenger cars per 10,000 people 2002	21 Cell phone subscribers as % of population 2002	22 Internet subscribers as % of population 2002	Provinces
Doctors per 100,000 people 2002	Expenditure on tobacco as % of expenditure on food 2002				
332	3%	512	65%	19%	Beijing
265	4%	202	30%	13%	Tianjin
144	6%	79	12%	4%	Hebei
199	7%	54	13%	4%	Shanxi
206	5%	55	15%	1%	Inner Mongolia AR
214	6%	52	21%	10%	Liaoning
210	4%	49	17%	4%	Jilin
169	4%	45	18%	4%	Heilongjiang
269	5%	88	56%	21%	Shanghai
139	6%	39	20%	4%	Jiangsu
161	7%	67	31%	7%	Zhejiang
97	8%	19	9%	2%	Anhui
129	4%	34	23%	7%	Fujian
114	5%	11	10%	3%	Jiangxi
144	3%	47	13%	3%	Shandong
106	5%	30	8%	2%	Henan
145	6%	21	10%	3%	Hubei
131	6%	23	10%	3%	Hunan
131	2%	101	41%	7%	Guangdong
107	2%	19	9%	2%	Guangxi AR
148	3%	18	14%	3%	Hainan
122	6%	19	14%	2%	Chongqing
134	6%	41	10%	1%	Sichuan
95	8%	13	6%	1%	Guizhou
119	9%	44	12%	2%	Yunnan
165	12%	39	8%	1%	Tibet AR
162	6%	37	13%	3%	Shaanxi
134	7%	21	8%	2%	Gansu
180	6%	40	15%	1%	Qinghai
171	6%	43	13%	3%	Ningxia AR
199	3%	68	16%	2%	Xinjiang AR
147	**5%**	**49**	**16%**	**4%**	**China**

Provinces	23 Water		24 Domestic water use liters per person per day		25 Industrial waste water discharged below standard 2002 million liters
	Renewable resources 2002 km³	Use as % of renewable resources 2002	Rural 2002	Urban 2002	
Beijing	1.7	203%	127	237	30
Tianjin	0.4	543%	93	176	6
Hebei	8.6	245%	67	202	878
Shanxi	7.9	73%	48	106	415
Inner Mongolia AR	31.5	57%	117	113	698
Liaoning	14.8	86%	87	181	1,118
Jilin	36.9	30%	76	169	800
Heilongjiang	63.3	40%	104	123	347
Shanghai	4.6	226%	105	406	334
Jiangsu	26.8	479%	116	248	1,072
Zhejiang	123.0	17%	112	304	618
Anhui	82.5	24%	51	153	275
Fujian	120.1	15%	127	217	342
Jiangxi	198.3	10%	97	203	1,033
Shandong	9.8	257%	67	131	387
Henan	31.4	69%	70	186	1,131
Hubei	115.5	21%	82	231	1,555
Hunan	256.7	12%	155	210	2,502
Guangdong	188.5	24%	162	450	1,501
Guangxi AR	237.3	13%	137	314	1,535
Hainan	33.3	13%	114	311	46
Chongqing	54.6	11%	93	254	850
Sichuan	206.6	10%	81	153	2,459
Guizhou	111.8	8%	90	226	740
Yunnan	230.9	6%	90	189	1,151
Tibet AR	424.3	1%	260	108	–
Shaanxi	25.5	31%	54	157	501
Ganxu	15.0	82%	57	192	546
Qinghai	55.8	5%	112	197	144
Ningxia AR	1.3	639%	44	180	507
Xinjiang AR	106.8	44%	174	317	524
					0
China	2,825	19%	94	219	24,149

CHINA: NATURAL RESOURCES 中

26 Energy production			27 Electricity consumption 2002 billion kwh	28 Industrial waste gas emission 2002 billion m³	Provinces
Coal production 2002 million tons	Crude oil production 2002 million tons	Natural gas production 2002 100 million m³			
9	–	–	44	297	Beijing
0	12.2	8.9	27	368	Tianjin
61	5.0	5.9	97	1,274	Hebei
244	0.0	2.1	63	940	Shanxi
89	0.0	8.5	32	600	Inner Mongolia AR
52	13.5	13.3	81	1,046	Liaoning
17	4.8	2.4	31	352	Jilin
59	50.3	20.2	47	463	Heilongjiang
0	0.5	4.3	65	744	Shanghai
26	1.6	0.2	125	1,429	Jiangsu
1	0.0	0.1	101	853	Zhejiang
61	0.0	–	39	512	Anhui
6	0.0	–	50	357	Fujian
14	0.0	–	25	261	Jiangxi
131	26.7	7.5	124	1,431	Shandong
99	5.7	19.4	92	1,065	Henan
4	0.8	0.9	56	644	Hubei
18	0.0	–	48	419	Hunan
2	12.6	31.6	169	1,058	Guangdong
5	0.0	–	36	569	Guangxi AR
0	0.0	–	5	53	Hainan
12	0.0	1.9	25	198	Chongqing
28	0.1	100.1	66	729	Sichuan
50	0.0	0.5	37	352	Guizhou
12	0.0	0.2	35	366	Yunnan
0	0.0	–		1	Tibet AR
59	10.6	40.0	36	342	Shaanxi
21	0.6	0.9	34	297	Ganxu
2	2.1	11.5	13	94	Qinghai
17	0.0	–	18	163	Ningxia AR
16	20.2	46.1	21	251	Xinjiang AR
1,380	**167.0**	**326.6**	**1,463**	**17,525**	**China**

NOTES AND SOURCES

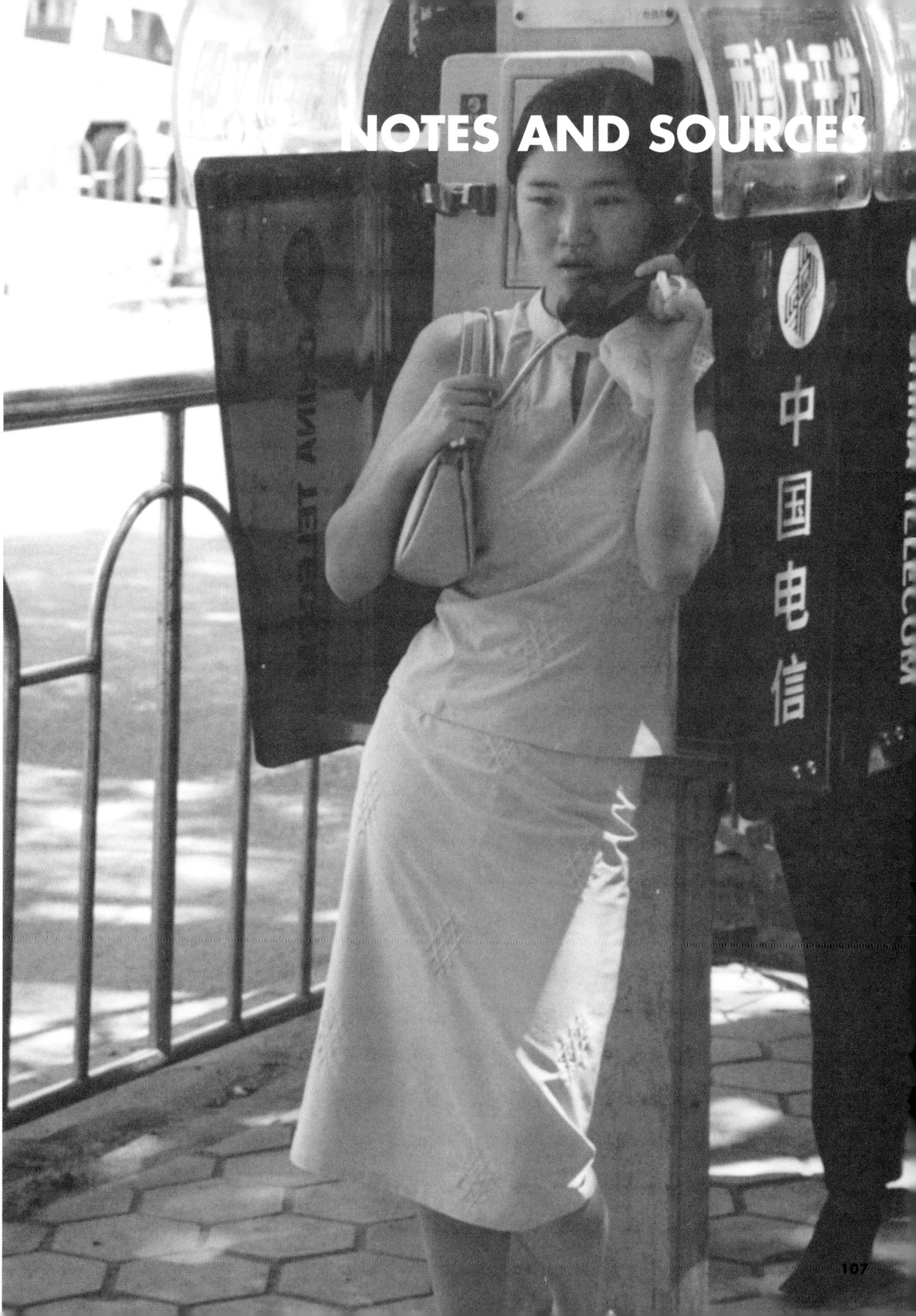

⊞ COMMENTARIES ON THE MAPS

The sources below each commentary include those for the text that appears on the map pages.

1 Trade

It will be many years before China challenges the USA's global military dominance. Trade, however, is another matter, as China is well underway to becoming a trading superpower. China's economy by 2010 is expected to be double that of Germany, which in 2004 was the third largest in the world. China is also expected to dislodge Japan as the world's second largest economy by 2020. Japan in 2004 imported more from China than from the USA. As a major trading nation, China is in a position to exert considerable power and influence in the global arena, but even more so in the region where the USA has reigned supreme since WWII.

There has been a remarkable growth in trade since China's open policy was declared in 1978 – a far cry from the previous Maoist program of striving for self-sufficiency. That policy was not without its success, for economic growth grew at an annual rate of 6 percent from 1952 to 1978. But foreign trade grew at a slower rate, with China's leaders fearful of losing control of the nation's industry to foreign powers. Another dramatic shift has been towards the export of manufactured goods. This cannot be credited solely to the availability of cheap labor, but also to sound economic management, which has led to striking improvements in the quality of Chinese goods and reliability of delivery, thus making them competitive on the world market. China's trade, however, is more than seeking markets for its products or investment in the economy; crucially, it also needs to import grain and oil, and to procure military and civilian high technology.

It is still too soon to measure with any degree of certainty the impact of China's accession to the World Trade Organization. There is the issue, however, as to what extent China can rely on trade as its springboard to world power. There are challenges and even threats to its domestic tranquility and stability, as related elsewhere in the atlas. These include a shaky financial and banking system, corruption, growing inequalities, unemployment and severe environmental problems. A growing pernicious nationalism may fuel hot-headed adventurism overseas. The strained relationships with the USA may erupt over Taiwan or, more likely, when America's status as the world's only superpower is challenged.

Sources: Barry Buzan and Rosemary Foot, *Does China Matter? A Reassessment*, London & New York, RoutledgeCurzon, 2004 • Alan Hunter and John Sexton, *Contemporary China*, Basingstoke: Macmillan Press, 1999 • National Development and Reform Commission, Survey, *People's Daily Online*, August 31, 2004 • Jason T. Shaplen and James Laney, "China Trades its way to Power", *New York Times*, July 12, 2004 • "China Embraces the World Market", Table 2A.3, Department of Foreign Affairs and Trade, Australia

2 Foreign Investment

Foreign direct investment (FDI) has been the motor of China's economic growth. China offers plentiful cheap labor, available land, minimal restrictions on working conditions, an authoritarian government committed to market-led reforms, and potentially a huge domestic market. In turn, foreign firms create employment, supply expertise and training, and bring technology. China gains economic growth and makes new friends. Business executives can be counted as among China's best allies. They persuade governments to restrain their criticisms of China's human rights abuses and lobby for export liberalization.

The impact of this partnership on China's economic development is staggering. In stark terms, China is the leading destination for global FDI, ahead of the USA. It accounts for approximately 20 percent of global FDI in developing countries. Cumulative FDI in the reform period between 1978 and 2003 exceeded US $400 billion. Actual, as distinct from contracted, FDI was 12 percent higher in the first half of 2004 than during the same period in 2003. The Asian Development Bank predicts that China will continue as the number one destination for FDI in 2005.

China's response has been the overdue investment in infrastructure. Foreign investors have responded by seeking control of their joint enterprises or by setting up wholly controlled subsidiaries. Foreign-funded enterprises were estimated to account for 31 percent of the country's industrial production in 2003 and more than half of its exports. In the pre-reform period this was what the Maoists feared and what stimulated their drive towards self-sufficiency. It is not difficult to imagine what they would think of the WTO. China's membership of the WTO allows foreign investors to enter new fields, including telecommunications, banking, insurance and the securities market. While attention focuses on China's ability to attract investment for both the domestic and export markets, it should be remembered that China is also a direct investor abroad.

The concentration of FDI in the eastern coastal provinces, as highlighted by the map on p. 61, has at least two important consequences. First, there has been an increase in the power of these provinces relative to the center, or, stated differently, there has been a *de facto* devolution of political power from central government to the provinces. Second, the disparities between the coastal and inland provinces are substantial. The Party-state has addressed the first through the reform of taxation policy and the second through its "Go West" program which initiates its own significant investment as well as attracting foreign investment for the central and western provinces. Perhaps more important is its ability to maintain political and social stability in order to attract and keep foreign investors. Reports of unrest and local conflicts in 2004, however disparate, were not reassuring for the investor. Unfettered growth is not an iron clad guarantee of political stability in an authoritarian Party-state.

Sources: Asian Development Bank, *Asian Development Outlook 2004 update, People's Republic of China* • Stuart Harris, "China in the Global Economy" in Barry Buzan and Rosemary Foot, *Does China Matter? A Reassessment*, London & New York: Routledge, 2004 • Shaun Breslin, "China in the Asian Economy" in Buzan and Foot, *ibid* • Peter Nolan, *China at the Crossroads*, Cambridge: Polity Press, 2004 • "Can Emerging Markets Follow China's FDI Growth Recipe", January 22, 2004 <www.weforum.org> • Edward Lanfranco, "Lash Slams China IPR Enforcement", August 12, 2004, UPI

3 Military Power

China is unlikely to come anywhere near challenging the USA's military dominance in the near future. In 2003 the USA's military expenditure was in a league of its own, over seven times more than China's. Yet China is determined to increase and prove its military capacity, and believes it has good reasons for doing so in regard to both its external and internal security needs. The military is also an important and powerful political actor, however wedded and beholden to the Party. The Fourth White Paper on Defense, issued in 2003, is clear about the PLA's priorities. The increase in military spending not only met the demands of the global war on terrorism, more advanced equipment and high technology weaponry, but salary increases and better conditions for PLA personnel as well.

The 11 major industrial groups that make up China's defense industry have been instructed to make a great technological leap forward to catch up with the USA. Inadvertently, the EU arms' sale embargo and the USA's pressure on Israel to drop its sales of radar equipment have forced China to develop its own systems and to do so quickly. Russia is the main alternative source of equipment and technology, and China is its largest customer. China, however, strives to achieve self-sufficiency in arms' production, so for every Russian system purchased there is a Chinese system being developed.

In increasing its contribution to the UN Peacekeeping Operations, China is also facing up to the international war on terrorism. The Fourth White Paper on Defense makes clear that support for China's international cooperation is conditional on acting through the UN, based on "conclusive evidence", "clear targets" and the "norms of international law". The Paper goes further in emphasizing the need to tackle the "root causes of terrorism".

Sources: John Hill, "China's Armed Forces set to Undergo Face-lift", *Jane's Intelligence Review*, vol 15 no. 2 February 2003, pp. 10–16 • David Isenberg, *Asia Times Online*, November 18, 2004 • IISS, *The Military Balance 2003–04* Oxford: Oxford University Press, 2004 • Dan Smith, *The State of War and Peace Atlas*, London: Earthscan 2003 • David Shambaugh, *Modernizing China's Military*, Berkeley and London: University of California Press, 2004

4 International Relations

China may not be a world power yet, but it is beginning to behave like one. As the world's fourth-largest economy, it has reason to do so. Moreover, the USA views China as a potential rival, and exerts pressure through both competition and control. As Nolan argues, China is in a different position in its international relations than were Japan, South Korea and Taiwan at comparable stages of development. They were close allies of the USA during the Cold War, and the USA protected them as developing states. When the Soviet Union collapsed and China opened up, the USA insisted that both follow the American model. The paradox is that as China sought acceptance as a member of the global community, the USA, under George W. Bush, withdrew towards unilateralism.

In China's international relations, history looms large, from greatness to humiliation at the hands of the imperialist powers, to the emergence from a half century of military struggle and chaos. China is bordered by 14 countries, and is engaged in a series of international political, economic and strategic relations, both bilateral and multilateral, which have developed unevenly. China has been integrated into the global community through its membership of international institutions, as a signatory to multilateral treaties and as a contributor to UN peacekeeping operations. Perhaps most important is its membership of the WTO. After 15 years of negotiations, and acceptance of many of the demands for reducing tariff barriers, China gained membership in December 2001.

China's development strategy of a market-led economy and political stability is driven by international trade. Acceptance into the WTO legitimates this. It also helps legitimate the WTO, and for that matter, other international institutions that cannot maintain their status and power without the membership of the world's largest country. The significance of China's membership depends on whether it meets it human rights obligations, and whether its decision makers can balance domestic modernization with international ambitions.

Sources: Rosemary Foot, *Rights Beyond Borders*, Oxford: OUP 2000 • Samuel S Kim, "China in World Politics", in Barry Buzan and Rosemary Foot, *Does China Matter? A Reassessment*, London & New York: RoutledgeCurzon, 2004 • Peter Nolan, *China at the Crossroads*, Cambridge: Polity Press, 2004 • Tony Saich, *Governance and Politics of China*, Basingstoke and New York: Palgrave Macmillan, 2004 • "President Jiang Zemin Met With Chilean President Lagos and Other Officials" <www.chineseconsulatevancouver.org> • "Chinese paper calls for closer China-EU" *English People's Daily Online,* October 14, 2004 • "Commentary: Anti-dumping Cases Use Flawed Data" *English People's Daily Online,* August 12, 2004 • "600 ballistic missiles" BBC News UK Edition, 27 October 2004 • South China Sea Region <www.eia.doe.gov/emeu/cabs/schina.html> • Alan Boyd, "South China Sea: Pact won't calm waters" July 21, 2003 *Asia Times* online

5 Population

In the 1990s when the first edition of the atlas was conceived, the population problem was still a metonymic shorthand for thinking about China. By 2004, the rate of population growth had slowed, although the actual population was still large, and still growing.

From the sheer weight of numbers, policy implications flow. In 2020, 11.7 percent of the population of China is expected to be aged over 65. Although this is a measure of the improvements in lifestyle, medicine and nutrition, it presents the problem of how China is going to deal with the pensions, housing and health care needed on such a massive scale. And, to complete the picture, the declining birth rate means there will be a smaller workforce to support a larger elderly population.

Uneven distribution of the population is a reflection of economic growth, and developmental contradictions emerge when the labor force is both the power and the problem of the reform era. China's leaders have pursued "a highly interventionist and aggressive policy" (Conway, 1997), whereby "population planning has shifted from being pronatalist in the 1950s to being anti-natalist" (Edwards, 2004). Population control was seen as an imperative to modernization, even though it conflicted with the belief in the 1980s and 1990s that it is China's cheap and plentiful labor force that makes it competitive in a global market. However, there is already high unemployment and, following its entry into the World Trade Organization, China faces shrinkage in heavy industries, and more lay-offs to manage among the workforce. The modernizing population has aspirational consumption patterns which support capital growth, and this will increase as urbanization progresses, but many people will be left out of the lifestyle explosion.

Furthermore, China's capacity to operate as the production house for many international concerns is likely over time to become less crucial than its ability to offer outsourcing for highly trained knowledge workers. Again, the population growth in the western regions may not be able to find work at all, whilst only those with access to good education and training in the east will do well – but they will then bear a burden of welfare for the aged and the unemployed.

Sources: Gordon Conway, *The Double Green Revolution*, London: Penguin Books, 1997 • UN Population Division statistics <http://esa.un.org/unpp/> • Louise Edwards, "Women's political work and 'women's work'", in Anne MacClaren ed. *Chinese Women: Living and Working*, London: RoutledgeCurzon, 2004

6 Urbanization

At the beginning of the reform period in 1978, 81 percent of China's population was classified as rural. Today, the proportion of the population that is rural is down to 62 percent, following the global trend towards urbanization.

Even where the population has grown in rural areas, this picture is complicated by the number of towns and small cities that have sprung up in those regions, and by the vast extent of domestic migration both within and across provincial boundaries. With the re-designation of towns as cities, and the economic status of cities on the rise in every province, the rural resident who does not have dealings in the urban environment is increasingly unusual. The flow of domestic migration is very great; it is estimated that several million (perhaps as many as 200 million) people are moving constantly between places of work and rural places of origin.

Rural job seekers are replacing urban workers by undercutting the market, and are causing long-term employment problems (Ma, 2004). Since the migrants are non-permanent residents, still in many cases without access to infrastructure and basic amenities, city populations are often underestimated – and under managed. Although migration does not necessarily entail movement to the biggest metropolitan districts, but often a lower scale of intra-provincial movement, the continuing intensity of population density on the eastern seaboard means that China's population remains unevenly spread across its territory (*p. 24*).

China's cities are most usefully understood as administrative centers with a very great concentration of people. City administrative districts can be visualized as in concentric circles: the old city/inner city; suburban districts which contain district-administered towns; commune/market towns. Then, outside the urban area, but within the municipality, there are rural counties that contain county-administered towns and more market towns and communes. Cities include the province-level cities of Beijing, Shanghai, Tianjin and, most recently, Chongqing, which is sometimes described as the largest city in the world; capitals of provinces or autonomous regions, or state-planning cities such as Dalian (Liaoning) and Qingdao (Shandong Province), which are responsible to the central government. For the purposes of investment and development some, such as Dalian and Qingdao, are also designated as Open Cities, and a smaller number, such as Shenzhen, are Special Economic Zones. The 2003 trade agreement (CEPA) between Hong Kong and mainland China encourages cross-investment and production. It is likely that such links will foster major urbanization, as the conurbation of Hong Kong's new territories leaps the border to stretch right across the Pearl Delta and Guangzhou.

Sources: Robert Benewick, "Towards a Developmental Theory of Constitutionalism: The Chinese Care", *Government and Opposition*, Autumn 1998 • Richard Kirkby, *Dilemmas of Urbanization: review and prospects*, London: Longman, 1994 • press reports, Ma Xiaohe, quoted from Tianze Institute report, "China reports 8% GDP growth in 2003", *People's Daily Online* <http://english.people.com.cn> accessed November 5, 2002 • Andrew M. Marton, *China's Spatial Economic Development: Regional Transformations in the Lower Yangzi Delta*, London: RoutledgeCurzon, 2000 • "China migration country study", Huang Ping (Institute of Sociology, Chinese Academy of Social Sciences, Beijing) and Frank N. Pieke (Institute for Chinese Studies, University of Oxford, UK) presented at the Conference on Migration, Development and Pro-Poor Policy Choices in Asia, Dhaka, 22–24 June 2003, published on internet • "First Impressions of the 2000 Census of China", November 4, 2001, William Lavely (Sociology Department and Center for Studies in Demography and Ecology University of Washington), published on internet • "China's urban population to reach 800 to 900 million by 2020", *People's Daily* September 17, 2004, cited on China Economic Net

7 The Gender Gap

Gender relations in China have been characterized by an historical contempt for women. Even in recent times, when the educational levels and employment achievements of the intellectual elites at least equals those of sisters in developed nations, women continue to be hampered by the priority given to men in social and cultural organizations. In rural areas, the situation is significantly worse, and recent analyses of the last census show a correlation between gendered work on the land and poor levels of education. This is happening particularly in those agricultural areas, which are otherwise described as developed, "because in the agricultural sector of these zones essentially women and older people are working" (National Bureau of Statistics, 2004).

Work is also a relevant way of plotting new class relationships in a modern society. Wanning Sun (2004) has shown that maids are ubiquitous in cities, many of them coming from rural areas in the same province, or further afield; Anhui and Sichuan are famous for their *baomu*. Now, however, as local state enterprises fail and the number of local unemployed increases, urban women are seeking the previously scorned role of housemaid, or nanny. At the other end of the spectrum, David Goodman (2004) has argued that a new entrepreneurial class of women – already seen in educated elites in Shanghai and Beijing, are also emerging in central provinces. He argues that despite the status of so-called "non-working wives" (*meigongzuo de laopo*), women are intrinsically involved with business, and their work is of central importance to the development of the family's entrepreneurial interests. It remains to be seen whether they will derive professional recognition or financial security from these activities. In the public sector workforce, women still tend to occupy informal, part-time, and low-wage positions. Women who appear in services statistics as "professional and technical", are generally nurses or primary teachers. These roles are respected, but carry less status and lower pay than many of the male-dominated professions.

Although the importance of family life to a healthy nation is a common thread running through the 20th century – in the social policy of the Imperial, Republican and Communist regimes – there have been exceptional periods. In the late 1960s girls were encouraged by political idealism to put off marriage and sex, and in the 1970s the birth-control program placed an emphasis on the way people behaved in relation to the rest of the population, rather than in relation to their own sexuality or desire for a family. This policy was called the later-longer-fewer (*wan-xi-shao*) plan. Women would marry later, have longer spaces between births, and fewer children overall. In 1979 the one-child policy was announced: a system of rewards and fines, designed to significantly decrease the proportion of children to adults. In 1981 the policy was tightened to demand that all families should restrict themselves to a single child until the year 2000. The plan was that the size of the population would stabilize in the interval, without having a long-term effect on the composition of society. Unfortunately, the fundamental problem of gender inequality has led to abusive practices against unwanted girl children, and latterly to abduction of prospective wives. Twenty-five years after the policy was introduced, it is reportedly under review.

Neither the Cultural Revolution of the 1960s, nor the delay of marriage in the 1970s had overturned the basic social assumption of male superiority, nor of women's imperative to be mothers and wives. There has been progress since Liberation (*jiefang*) in 1949, however. Women are now involved in public life, especially at local levels. The number of female deputies in the National People's Congress (Map 19) increased from 147 in 1954 to 626 in 1993, and has remained steady at approximately 21 percent of delegates in subsequent meetings. Many political women are, however, engaged in, or assigned to "women's issues", a situation which exacerbates gender divisions in the sphere of government. Nonetheless, the Chinese Women's Ninth National Congress (2003) reported grassroots work on significant issues, including literacy, domestic violence and the establishment of rural-women's schools to create "farmer technicians". These initiatives are vital. School enrolments are lower for girls than boys, especially in rural areas. About 72 percent of illiterate people in China are women.

The future of Chinese women is mixed. If the gender gap continues to widen there will be more violence against women as they become a scarce, but undervalued, "human commodity" in an increasingly market-oriented society. Some women will benefit from their single-child status, but this is not a significant factor, as intellectual and professional families have long taken the education of their daughters very seriously. Women will suffer if the employment rate drops, and men demand access to jobs in a shrinking field of opportunity. The gender gap is a symptom rather than a cause of women's struggles for equal status with men in public, economic and domestic life in China today. Women will need to increase their role in the building of governmental systems, in accessing economic power in the era of reform, and in avoiding the worst degradations of capitalism on their bodies and self-respect.

Sources: Colin Mackerras, *The New Cambridge Handbook of Contemporary China*, Cambridge: Cambridge University Press, 2001 •Tamara Jacka, *Women's Work in China: Change and Continuity in an Era of Reform*, Cambridge and Melbourne: Cambridge University Press, 1997 • Penny Kane, in "Population and Family Policies", Benewick and Wingrove (eds), *China in the 1990s*, Basingstoke: Macmillan (revised 1998) • *Far Eastern Economic Review*, various issues • David Goodman, "Why Women Count: Chinese Women and Leadership of Reform"; Wanning Sun, "The Maid in China: Opportunities, Challenges and the Story of Becoming Modern" both in Anne E. McLaren ed. *Chinese Women – Living and Working*, London: RoutledgeCurzon, 2004 • National Bureau of Statistics, <http://www.stats.gov.cn/english>

8 Minority Nationalities

Although minority nationalities in China make up a small proportion of the population, because of the size of the overall population they comprise a very large number of people. Just one minority group, the Zhuang, are more numerous than the current population of Chile, and the minority nationalities as a whole outnumber the population of Australia. The term, "minority nationality" (*zhongguo shaoshu minzu*), is based on a Stalinist definition of nationality: "an historically constituted community of people having a common territory, a common language, a common economic life, and a common psychological make-up which expresses itself in a common culture." (Wang, 1998) In theory, this definition works in favour of political citizenship, against a pan-Chinese ethnic identity. In practice, there are documented economic and social disadvantages attached to minority status for many groups.

Recognition of minority status currently applies to 56 groups, including the dominant Han majority. Despite the insistence on discrete national identities, there are many similar

elements within the experience of many minority groups, whatever their "nationality". These can be summed up as: economic disadvantage, poor representation at national levels, religious repression and educational disadvantage. Direct repression is only encountered when the State perceives religion operating as a galvanizing force in demands for independence (as opposed to autonomy).

The most highly publicized case is that of Tibetan Buddhism. However, the situation in Xinjiang is, if anything, more troubling to central government. Xinjiang is known to Muslims in Central Asia as East Turkestan, a name that relates the predominantly Muslim Uygur population to their neighbours in the newly independent states of Kyrgyzstan, Kazakhstan, Uzbekistan, Tajikistan, and Turkmenistan. These were formerly part of the Soviet Union, but are recalled in the memory of their indigenous peoples as parts of a greater Turkestan, which converted to Islam 1,200 years ago. Unrest was reported in Xinjiang in the 1990s. Since 2001 the central government has grown more repressive and more nervous of the aspirations of the western peoples. It is arguable that the problems may be as much to do with central government containment strategies as with long-standing religious allegiance.

Large-scale migration of Han and other ethnic groups into Tibet and Xinjiang has produced a two-tier economic system, with the Tibetans and the Uygurs sitting at the bottom of the economic ladder in their own areas. Many of the incomers have been troops, stationed semi-permanently to reclaim wasteland and establish new oases. This policy, which echoes age-old centralist ideas of "strengthening frontiers through people" (*yimin shibian*), is dangerous to fragile environments but is also perilous for the people who find themselves washed osmotically out of the structure of development by better-educated and more highly skilled migrants. By 2000, 41 percent of inhabitants of Xinjiang were Han.

The dual-structure of the economy of Xinjiang, developed through Han and other ethnic immigration, is structured around modern industry in the north, and an agricultural base in the south. The Uygurs tend to reside in the poorer south, whereas recent migrant populations operate and expand in the north. There have been similar experiences in Tibet and Inner Mongolia, although not as obviously as in Xinjiang. Between 1964 and 1994, 70 percent of migrants into Tibet were themselves Tibetans from nearby provinces. Migrants into Inner Mongolia have been predominantly Han.

Key indicators still suggest economic disadvantage for minorities: life expectancy in Xinjiang is 67.41 years – compared with a national level of 71.40 years. This makes it the fifth lowest after Qinghai, Guizhou, Yunnan, and Tibet, all areas where over 25 percent of the population are from national minorities.

Sources: Robert Barnette, ed., *Resistance and Reform in* Tibet, London: Hurst, 1994 • Colin Mackerras, *The New Cambridge Handbook of Contemporary China*, Chapter 9, Cambridge: Cambridge University Press, 2001• David Wang, "Han Migration and Social Changes in Xinjiang," *Issues and Studies*, vol. 34, no.7, July 1998, pp.33–61 • Zhang Tianlu Huang Rongqing, *Zhongguo shaoshu minzu renkou diaocha yanjiu* (*Surveys and Research into China's Minority Populations*), March 1996, Gaodeng Jiaoyu Chubanshe • Amir Teheri, "The Chinese Muslims of Xinjiang," *Arab View* <http://www.arab.net/> • "Promoting Three Basic Freedoms: Freedom of Association Assembly and Expression", September 1997 <http://www.igc.apc.org/hric/> • press sources.

9 Entrepreneurial China

The new face of state capitalism in China, and the emphasis on creating a market economy, has seen a rapid increase in private enterprise, and in the number of entrepreneurs. The growing entrepreneurial culture is also a symptom of improving ties between mainland China and "Greater China" – the term used for Chinese populations in Hong Kong and Southeast Asia that have a developed commercial connection with the People's Republic. Whatever political differences are at play, Chinese populations continue to consider one another as natural and viable partners in business and trade. These relationships have matured over the past ten years, and have been encouraged by China's accession to the WTO, and by the more local CEPA agreement between Hong Kong and the Mainland – generally understood as a local Pearl Delta initiative with Guangdong. The idea of "Greater China" is regional and strategic, frequently drawing together a triangle of southern China, Taiwan and Hong Kong–Macau. This informal grouping is developing in parallel with international organizations, such as the Malaysian-led EAEC (East Asian Economic Caucus), ASEAN (which excludes Taiwan but is traditionally anti-Communist), and APEC (Asia-Pacific Economic Cooperation), which incorporates non-Asian countries into a forum for inter-regional co-operation.

Entrepreneurialism at home responds to these large opportunities from overseas and to domestic conditions. State entrepreneurialism is evident at city and provincial levels, and also across certain sectors, such as education, advertising, real-estate development, and tourism. The Shanghai Television University, for example, is pushing distance and online learning in order to capture the vocational market from its rivals in the tertiary sector, and other television channels. As Jane Duckett has pointed out, state entrepreneurialism is likely to grow with the availability of markets, and the improvement in bankruptcy laws and loans systems. A political will to devolve economic micro-management allows state actors to take more responsibilities and more risks in their sector or region.

Individual entrepreneurs are operating within a similar set of factors. The de-collectivization of state assets, the 2004 decision to support private land ownership, increasing unemployment, and the need to look after one's own and one's family's welfare in old age, is pushing people into ambitious ventures. Again, quoting Duckett: "Entrepreneurialism may have spread because of perceptions that everyone else is doing business, and those who do not feel that they are missing opportunities."

Sources J. H. Chang, P. Kee and J. Chang (eds), *Chinese Cultures in the Diaspora: Emerging Global Perspectives on the Centre and the Periphery National Endowment for Culture and the Arts*, Taipei, 1997 • Ankie Hoogvelt, *Globalisation and the Postcolonial World: The New Political Economy of Development*, Basingstoke: Macmillan, 1997 • Stephanie Hemelryk Donald, Terry Flew, and Wang Xianchao "The Administration of Creativity: China. Leadership and the new MBA" (unpublished) • J. Duckett, *The Entrepreneurial State in China*, London: Routledge, 1998 • D. S. G. Goodman, "Why Women Count: Chinese women and the leadership of reform", in Anne E. Mclaren ed., *Chinese Women: Living and Working*, RoutledgeCurzon: London, 2004 • Liu Jie, "Property Prices Increasing Rapidly", *China Daily*, August 20, 2004

10 Equality and Inequality

The people of China experience very different life-styles, according to the region in which they live, whether they are urban or rural, whether they are male or female, and depending on their access to the infrastructure necessary for economic development. Although poverty decreased until 2003, the trends are not good, and the gap between rich and poor is apparent and increasing. Some of these differences are historical and geographical. The Western Provinces are rich in mineral resources and space, but historically poor in political influence and basic infrastructure. The coastal provinces have easy access to trade by sea, a strongly developed urban network, and close connections with the seat of power in Beijing.

Some provinces look decidedly disadvantaged when compared with other, more prosperous, provinces. Large, predominantly rural, provinces, such as Qinghai, Gansu, Yunnan, and Guizhou are disadvantaged by their distance from the center, by their low proportion of successful commercial activities, and by the slower rate of modernization they experience. The government is working hard to develop these regions, and the increase in telephone subscribers in rural areas is an indicator of this resolve (*p. 81*). Guangxi, in the southwest, is also isolated, and has suffered badly from floods (*p. 81*), high inflation and a decline in rural income levels. Although it is also receiving special attention, many areas remain undeveloped and inaccessible.

Many of the poorer regions in China are populated by minority nationalities (*pp. 30–31*). This must give the Chinese government cause for concern. If the economic revolution continues, and the casualties of the market continue to increase, there will be a greater likelihood of disaffection amongst all rural communities, but particularly amongst those in minority areas.

One solution is to devolve financial and fiscal powers to the provinces (*fangquan rangli*) in the hope of controlling resources, containing unrest, and developing economic links across China that are perceived to be equal and mutually profitable. The distribution of wealth across the provinces is complicated by the large income differentials within provinces. Rural counties, even those adjacent to one another, are experiencing varying levels of economic advantage. This may be another incentive to increasing provincial autonomy, which goes hand in hand with existing attempts to "match up" rich and poor provinces in an effort at poverty alleviation. This is known as "horizontal regional co-operation", (*hengxiang jingji lianxi*). All such plans, as well as fundamental attempts to reduce illiteracy and safeguard welfare provision (*pp. 74–75*), are reliant on the continuing solvency of the nation's economy.

Sources: John Gittings, *Real China: From Cannibalism to Karaoke*, New York: Simon and Schuster, 1996 • Diana Hwei-an tsai "Regional Inequality and Financial Decentralization in Mainland China" *Issues and Studies*, May 1996, pp. 40–71 • B. Andreosso-O'Callaghan and Wei Qian, "The PRC's Economy - From Fragmentation to Harmonization?", unpublished MS, 1997 • Christina P. W. Wong, "Central–Local Relations in an Era of Fiscal decline: The Paradox of Fiscal Decentralization in Post-Mao China", *The China Quarterly*, Dec 1991, no. 128, pp. 691–715 • Ya-chun Chang, "The Financial Autonomy of Provincial Governments in Mainland China and Its Effects", *Issues and Studies*, vol. 32, no. 3, March 1996, pp. 78–95 • Jae Ho Chung, "Beijing Confronting the Provinces", *China Information*, vol. 9, nos 2/3 (Winter 1994/5), pp. 1–23 • Feng-cheng Fu and Chi-keng Li, "Disparities in Mainland China's Regional Economic Development and their Implications for Central–Local Economic Relations", *Issues and Studies*, vol. 32, no.11, November 1996, pp. 1–30 • Alan P. Liu, "Beijing and the Provinces: Different Constructions of National Development", *Issues and Studies*, vol. 32, no. 8, August 1996, pp. 28–53 • Joseph Fewsmith, "The Political and Social Implications of China's Accession to the WTO", *The China Quarterly*, 2001

11 Employment

The nature and requirements of work in China are very different from the years of peasant agriculture and heavy industry, supplemented by political and intellectual work. People have re-invented themselves as migrant laborers, service workers, small-scale entrepreneurs, and business managers. These re-inventions are in many cases linked to higher incomes, and more ambitious career options, and an increase in professional management posts. Many university graduates are still assigned to public posts in the state sector. However, in 2001, a total of 12,173 people enrolled for MBA courses in China, indicating that continuing employment is now seen as a personal responsibility. Even public servants need to think about their entrepreneurial skills in order to keep their jobs. An educational television boss in Shanghai commented in 2004 that the pressure (*yazi*) of his work was so great that he could think of nothing apart from how to turn "programming into profit".

There are several factors contributing to new employment profiles in China: reform and modernization, recent history, and regional imbalance in capital distribution and labor costs. Unemployment surfaced in the late 1970s when tens of thousands of people in their twenties and early thirties returned to the cities from the countryside. These were "urblings", city youth who had spent much of their life since the age of 17 working in remote rural areas as part of the management of the Cultural Revolution. Their impact on employment figures were, however, minimal in comparison with the influx of migrant peasants which they foreshadowed. And the effects of surplus labor continue. The state-owned enterprises sector needs to lay off significant numbers of workers if China is to meet its 2005 commitments under the World Trade Organization agreement. Some of this short-term pain is being re-directed to the private sector. The policy of "holding onto the big ones but letting the little ones go" (*zhuada, fangxiao*) ensures that only strategic, larger enterprises are kept under the wing of total state ownership.

There remain deep regional divisions. Employment patterns in the coastal regions are capital intensive and highly mechanized; laborers supplement their incomes with alternative and seasonal employment. Meanwhile, farm practices in remote regions are labor intensive but generate very low wages. Family members, often young women, need to move into towns and cities as guest workers to supplement the household income. They achieve a certain status through their shift into economic power, but their situation in the urban centers is often precarious. Development in industrial practices often, at least in the short term, leads to downscaling, re-organization, and the subsequent displacement and reconfiguration of the workforce. The worst-hit areas are in the central and western regions, especially Wuhan and the enormous Chongqing, which sustains a population of over 31 million.

Women workers are particularly disadvantaged by the likelihood of unemployment. Over half of newly laid-off workers are women, although they make up only one-third of the workforce. This adds to an already weak position for women, as only 38 percent were working for wages in 2002. They are further challenged by the illegal, cheap labor on hand from poorer regions. The choice is being made between social problems and a strong national economy. It is still uncertain which way the state will move in the long term.

Sources: Yunhua Liu, "Labor Absorption in China's Township and Village Enterprises", Paper for the International Conference of Economics of Greater China, 1997 • "Zhongguo xinwen she", China News Agency, May 2, 1997 • Fang Shan, "Unemployment in Mainland China: Current Situation and Possible Trends," *Issues and Studies*, vol. 32, no.10, October 1996 • "Lateline", ABC Australia, March16,1999 • John Gittings, "China puts its faith in enterprise", *Guardian Weekly*, March 21, 1999, p. 7 • Jean-Louis Rocca, "Unemployment sweeps China", *Le Monde Diplomatique* (International edition), February 1999, pp. 6–8 • Lina Song, "The Determinants of Female Labor Migration in China: a case study of Handan", Institute of Economics and Statistics, Oxford University; paper presented SOAS; June 8, 1995 •Tamara Jacka, "Working Sisters Answer Back: The Presentation and Self-Presentation of Women in China's Floating Population", *China Information*, vol. 13 no. 1, Summer 1998, pp. 43–73 • Peter C.Y. Chow, "China's sustainable development in global perspective", *Journal of Asian and African Studies*, vol.38, issues 4–5, 2003 •Wu Shinong, Tong Yunheng, eds. *MBA Program in China*, Beijing: Machine Industry Publisher, 2001 • *China Statistical Yearbook*, China Statistical Publishing House,1998 and 2004 • *People's Daily*, August 26, 2004

12 Agriculture

China is still a rural society, but links between town and village are increasing, as seasonal migration, small enterprise and urbanization of local townships change the expectations and challenges for the Chinese peasantry. Rural enterprises vary in size and scope. In the 1980s and early 1990s, the town and village enterprises were mainly collective, but there are now many more individual and partnership concerns, some involving overseas and urban investors. Rural workers who lack capital to join in with diversification are not advantaged by its success. In 2002, agricultural incomes per capita were 35 percent of those in the city.

It is a priority to modernize safely, but the pressures are great in a period of uneven development. State investigations have demonstrated that 77.6 percent of pesticides are safe, according to current standards of acceptability, but there remain large quantities of unsafe chemicals, mostly due to labelling problems, substandard formulas, complicated distribution channels, and overuse of chemical sprays as a quick-fix solution to productivity.

The land itself is under threat from short-term money-spinning development. The problem was addressed by the 1997 promulgation to the effect that all agricultural land conversion was frozen for 12 months, but that has been overturned by the WTO and its implicit support for cash crops. Productivity is higher in coastal areas where the land has been turned over to new crops, many of which are exported to Japan – which purchased over 50 percent of vegetables and fruit products between 1998 and 2000.

In 2004, the government introduced measures to boost grain output, which had been declining year on year since 1999, and was no longer meeting China's needs. 10 billion yuan was earmarked to compensate farmers for losses caused by low grain prices, and 1 percent was knocked off agricultural tax rates. Financial incentives were offered for the purchase of farm equipment, or the digging of wells for irrigation. The early signs were that they were having some effect in boosting production, although a comparison with the previous year's drought-affected harvest gave an unduly rosy picture. From the global perspective, China's grain production is of some significance. If the country were to become a major grain importer it would be likely to have an inflationary effect on the world grain market, and could cause grain shortages elsewhere. From a national point of view, food security needs to be maintained in case of international crisis. At the same time, China needs to grow its labor-intensive, higher-value crop yields to support its trade and to maintain stability in the countryside.

Sources: Jane Sayers, Environment Victoria • Fang Shan "Unemployment in Mainland China: Current Situation and Possible Trends", *Issues and Studies*, vol. 32, no. 10, October 1996 • "Agricultural Pesticide Use in China", "Irrigation in China Demands More Efficient Technologies <http://www.redfish.com/USEmbassy-China/sandt> • "China's government helps boost grain output" <http://english.people.com.cn> • East Asia Analytical Unit, DFAT, Australia • US Department of Agriculture, 2002.

13 Industry

The industrial sector in China is rapidly diversifying. On the one hand there are still a large number of state-owned enterprises (SOEs), many with bad debts and production problems These industries have traditionally underemployed, and have in the past underwritten their workers' welfare from cradle to grave. Government services have been delivered through the auspices and institutions of industry.

Meanwhile, private enterprises are being actively encouraged, and may soon have access to state bank loans. In some counties "attached" businesses, those which operate privately but are formally registered as public companies, are moving into an openly private classification. The mood is very much in favor of the professional entrepreneur, although some support is given to state industries in acknowledgment of the work they do for welfare, if not for national solvency. Successful entrepreneurs tend to be young, flexible, willing to travel, and able to free themselves of the restraints of "connections" (*guanxi*). By contrast, corruption in small manufacturing industries is often linked to a hierarchy of entrenched political interests, and a whole web of immovable links and favors, rotting the system.

The danger of burgeoning private enterprise, but also of public enterprises under pressure to reform and perform, is very real. Safety in the workplace is not good. In 2002, there were 2,765 deaths in coal mines alone. And, whilst state-owned enterprises reduced their fatalities by 2.2 percent, smaller work places became more dangerous – or rather their proliferation is leading to unregulated activity and irresponsible work practices. In total in 2002, there were 5,791 deaths in industrial accidents, 4,031 of those in town and village enterprises. Migrant workers (mingong) are most at risk in these situations. Many of them live in hostels run or leased by the factory, but with none of the state benefits administered by the SOEs. There is also the phenomenon of cross-border enterprises from Hong Kong. In the 2003 typhoon, nine seasonal workers died in Guangdong, on a building site financed by Hong Kong investment.

Sources: Linda Wong and Ka-Ho Mok, "Dynamism and Development: Economic Growth and Social Change in Post-Mao China", *The Asian Journal of Business Administration*, vol. 18, no. 2, December 1996, pp. 201–33 • Jonathan Unger and Anita Chan, "Inheritors of the Boom: Private Enterprise and the Role of Local Government in a Rural South China Township," Murdoch University Asia Research Centre Working Paper 89, 1999 • Jonathan Unger, "The Rise of Private Enterprise in a Rural Chinese District," Murdoch University Asia Research Centre Working Paper 90, 1999 • *South China Morning Post* various dates October 2004 • State Administration of Workplace Safety Online, 2003 • G. Jefferson *et al* "Ownership, performance and innovation in China's large and medium-size industrial enterprise sector", *China Economic Review*, 2003, 14–1, pp. 89–114

14 Services

The service sector was a political non-starter in the years before the reform era of the 1980s and 1990s. Maoist perspectives emphasized production, and the only valid services were those provided by government agencies. The re-growth of service industries, particularly in southern China, has moved in fits and starts since the 1980s, and with varying degrees of transparency. Department stores, luxury goods, upmarket restaurants, and late-night clubs are features of most cities. Real-estate, information and finance markets, retail and personal services are all growing. There is even an "English" village, complete with retail outlets from major British chains, planned for the suburbs of Shanghai.

There is also some overlap with what is known as the "quaternary sector" – digital and information-based services across the economy. Plans for the Shanghai Infoport are a prime example of such developments. Construction of the project started in 1996 and it is scheduled for completion by 2005. The infoport is designed to provide a complete information network and set of social resources for Shanghai, which will allow it to compete with world cities and world finance sectors. This is also dependent, of course, on the management of finance and banking for an open system of international exchange

Unplanned and statistically invisible services are also on the rise. The domestic sector is booming, with most professional women with children employing a nanny, maid or *baonu* to do the domestic chores. Hawkers are also common on city streets, but they need to be licensed. Those who come in from the countryside to sell off surplus produce have operated under "specialized agricultural households" licenses. Those selling food or manufactured produce on city streets are also required to obtain "temporary business licences". The penalty for non-compliance can be fairly straightforward. Police are quite likely to throw large piles of unlicensed produce into the canal.

Sources: Stephanie Hemelryk Donald, personal interviews • Lynn T. White III, *Unstately Power: Local Causes of China's Economic Reforms*, New York: M. E. Sharpe, 1998.

15 Tourism

The tourist industry is a major non-productive growth area in the Chinese service sector. Although pilgrimages and festival days have promoted internal travel across most periods of Chinese history, modern tourism is a recent development. Nevertheless, established private companies are already experienced enough to refine their developments in line with local and international expectations. Some companies are centered on villages lucky enough to be situated near a genuine "site". Villages in rural Beijing at the foot of the Great Wall moved fast in the 1980s to develop service facilities (and hawking) to maximize the benefit of their proximity. Between 1949 and 1972 the wall was only accessible to "special friends of the Revolution" and only 250,000 visas were issued in those years. The wall is now a necessary part of any Beijing visitor's itinerary. The phenomenon of local site development is now happening all over China. Guizhou, just one of many possible examples, has invented traditions to bring together its real and imagined minority cultures into a package for tourists.

Companies are promoting tourism by exploiting natural resources. Cave formations attract visitors to areas around Hangzhou and Guangzhou. As tourist sites are priced and licensed according to the number of attractions presented, cave management companies build wonderlands of lights, mythological theme parks, laser shows, and underground river trips into their programs. Developments in the early

1980s tended to be technologically simple. Successful companies are now re-investing in high-tech light shows and amenities to remain competitive in a fierce market. In 1993 rival villages went into battle to protect their claims to local caves, such was the projected value of the tourist trade.

Despite the long-running international discomfort with China's presence in Tibet, there is a strong focus on the region to attract domestic (Han Chinese), and some international, tourists. Lhasa is becoming the fulcrum of tourist routes across the plateau, and the region is investing heavily in rail, roads and facilities to continue the trend.

Hong Kong has long enjoyed the status of prime destination in the region, but in recent years the city has been struggling to retain that pre-eminence. Strong branding campaigns to promote Hong Kong as "Asia's World City" have been directed both at international and mainland markets since 2000. In October 2004, 164,000 mainland visitors travelled to Hong Kong in the first three days of the Golden Week holiday. Golden Week is a vacation break invented to coincide with the lunar festival, and to encourage domestic tourism around China. Hong Kong's controversial West Kowloon and harbor reclaims are part of efforts to beautify the city for visitors, and Shanghai has similar plans. In 2004 an urban planning exhibition promised: "a tourist system composed of downtown shopping, community activities and recreational facilities in the outer suburbs."

Sources: Tim Oakes, Tourism and Modernity in China. London: Routledge, 1998 • Sarah F. M. Li and Trevor H. B. Sofield, "Tourism development and socio-cultural change in rural China", A. V. Seaton (ed.), Tourism: The State of the Art, Chichester: John Wiley & Son, 1994 • Sarah F. M. Li and Trevor H. B. Sofield "Is The Great Wall of China The Great Wall of China?" Paper given at the 14th World Congress of Sociology. Montreal, 1998 • Zhang Zhiping, "Exploration Tourism: A New Pursuit", Beijing Review, November 24–30, 1997, pp14–16 • " World Expo and the new development in Shanghai: 2010 Expo and Shanghai Tourism", Shanghai: Shanghai Tourism and Enterprise Committee, Policy and Regulation Section, p.12, 2003

16 Traffic

In 1952, 67 percent of travellers in China journeyed long distances by rail. That had fallen to 7 percent by 2002 (although with only a fractional decrease in actual numbers travelling, owing to the massive increase in total population). Over the same period, highway traffic took up the dominant position, moving from 19 percent to 92 percent of domestic volume.

Motorized transport in China is still predominantly communal – buses and taxis are much more common than private cars – but that situation is changing rapidly. There is also a vast regional variation in the provision of public transport. In Beijing in 2002 there were 823,800 buses, cars and trucks on the roads. In Henan, a province with approximately seven times the population of Beijing, there were only 691,500.

These figures suggest that the rather hackneyed Western image of China – still trotted out in advertising campaigns – of bicycles clogging the city streets, is only partly true. What the advertisements do not show is the diversity of vehicles – from antiquated diesel trucks, to a few horse-drawn, and man-drawn, carts, to various colours, shapes and sizes of taxi-cabs – which share frantic city traffic jams with the bicycles. It is arguably this diversity that causes many of the road traffic accidents in major cities, and makes travelling by road in the countryside and the city outskirts so hazardous. There were 104,372 deaths attributed to traffic accidents in 2003. (Only just over half the per capita death rate in the USA, but an alarmingly high total, given the relatively few vehicles on the roads.) Meanwhile, urban planners in Shanghai and Beijing, where traffic is far less diverse – mainly bikes, taxis, buses and private cars – are looking to accommodate more private cars, rather than move towards long-term sustainability. On the bright side, however, Shanghai is also building a public-transport hub to support the 2010 World Expo, and Beijing has recently completed a new subway line, which encompasses the city's northwest university district, Haidian.

Long-distance travel is crucial to the opening up of the Western Provinces and to economic restructuring in general. The rail link to Tibet is a case in point, although a sore point for proponents of less Han influence and more independence. Air travel is increasing sharply, in response to intra-provincial business opportunities, domestic tourism and international travel. The lower-paid workers, described as domestic migrants, will, however, still be more likely to travel by train.

Sources: "The Road to Progress", Asiaweek, May 9, 1997, p.10 • China Statistical Yearbook, 1997 and 2005 • Thanks to Jane Sayers (Environment Victoria); Guo Yang (Shanghai Tourism Commission); ISTP, Murdoch University • World Health Organization press release announcing publication of World Report for Road Traffic Injury Prevention, October 8, 2004 • BBC News website, September 3, 2004

17 Energy

China produces 12 percent of the world's energy, and is the largest coal producer in the world, accounting for 37 percent of total production. Around 80 percent of the coal is produced in the north and the west, far away from the rich, industrial, coastal provinces. Much of it is of secondary quality, produced in small, inefficient and dangerous mines. The brown coal of past smogs and present respiratory illness is a target of the Kyoto Protocol. In addition to the environmental and health costs, coal contributes to the transportation bottleneck, since it takes up more than 40 percent of rail freight and is shipped by river and sea as well.

The crucial question for China is whether its energy resources are sufficient to meet its growing needs. Natural gas, which makes up 3 percent of China's energy production, is located in the remoter parts of China. A recent deal with Western Australia will also provide an offshore supply. Oil makes a major contribution to the country's energy needs, but despite being the sixth-largest oil producer in the world, China is now a net importer. Both off-shore production and oil-field explorations in Xinjiang province have proved disappointing, and China has resorted to buying oil fields in countries such as Azerbaijan and Indonesia.

The image of Shanghai's skyline, sparkling with light and future promise, needs to be taken with a note of pessimism. The lighting up of the sky is unsustainable without a change of attitude to consumption. China's electricity supply is groaning under the strain imposed by rocketing demand, which in early 2004 was 14 percent up on the previous a year. A campaign has been launched to get people to cap air conditioning at 79 degrees Fahrenheit (26 degrees Celsius), and other restrictions will be necessary in the near future.

Up until the mid-1990s China had an electricity surplus, and when the government closed down some of its inefficient, electricity-guzzling companies in the late-1990s it expected the planned power supply to meet demand for the foreseeable future. However, it failed to anticipate the rapid growth in China's economy, which meant that by mid-2002 there was an electricity shortage. Although at least 30 major new power projects were approved during 2003, experts estimate that it will not be until 2006 that the problem is eased. In the meantime, companies building new manufacturing or industrial plant in China need to consider providing their own source of electricity.

Nuclear power comprised 2.2 percent of total energy output in 2003, since when further reactors have been started up, and yet more are under construction or in the planning stage. Hydroelectric power is also significant – another factor in the relentless dam construction in China.

Sources: "Slick moves for oil expansion", *China Daily*, January 29, 2002 <www.chinadaily.com.cn> • Jayanthi Iyengar, "China power crisis dims production", *Asia Times*, September 24, 2004 <www.atimes.com> • World Nuclear Association, "Nuclear Power in China", September 2004 <www.world-nuclear.org>

18 Chinese Communist Party

China is governed by a relatively small number of leaders, who exercise power, both formally through a multiplicity of power structures, and informally through a network of contacts. Viewed from this perspective, China does not differ from most nation-states. Yet it is radically different. As the most powerful of the few surviving communist states, the Party leadership refuses to entertain the possibility of a legitimated and institutionalized opposition. Even the Democratic Parties remain under the leadership of the Communist Party. One of the legacies of the crackdown of the Tiananmen protest of 1989 is that the Party-state served notice that it would severely punish unauthorized mass movements and public demonstrations.

China's formal power structures include, first and foremost, the Communist Party. Other power structures include the government, the bureaucracy, the PLA, the judicial system including the police, and the provinces individually and collectively, but all of these power structures, although separate, are pre-eminently related to the Party. The structures of the Party parallel and penetrate those of the government, with Party Leading Groups functioning at all levels of the state, and there is an overlap of personnel. The Party contends and proposes, the state amends and disposes.

The formal power structure, how power is exercised informally, and who exercises it can all be described in terms of a pyramid structure. This pyramid structure is also evident in the organization of the Party. There is a hierarchical relationship, concentrating power at the apex of the pyramid and exercising control down to the base. Democratic centralism may be the form providing the opportunities and avenues for debate and discussion at all levels of the Party to be communicated upwards, but decisions are transmitted downwards.

Rather than presenting the National People's Congress as the highest decision-making body (p. 56), as described in the constitution, a representation of power (p. 57) shows that decision-making is concentrated in the Standing Committee of the Politburo, who rule on behalf of the Party. An analysis of the membership of the Standing Committee (p. 55) demonstrates how power has passed from a generation of revolutionary and veteran leaders to a generation of technocratic leaders. All nine members of the Standing Committee hold engineering degrees, all but one from Chinese universities. The analysis also reveals that six of the nine members simultaneously hold the highest-ranking positions in the government and are at the apex of the power pyramids. Hu Jintao is General Secretary of the Party, President of China and Chairman of the Central Military Commission; Wen Jiabao is Premier; and Wu Bangguo is Chairman of the Standing Committee of the National People's Congress.

Power and its exercise, however, is not always visible, and in China much depends on the base of support, and personal and institutional backing. Although Jiang Zemin has retired as Party Secretary and State President he still wields influence if not power through his followers in well-placed positions. Deng Xiaoping was the most powerful leader in China even when he held no official position. In popular esteem, he was the paramount leader.

Sources: There is no general authority on the Communist Party but good analyses are found in Kenneth Lieberthal, *Governing China*, New York and London: Palgrave Macmillan, second edition, 2004; Yun-Han Chu *et al* (eds.), *The New China*, Norton: 1995, 2004; Tony Saich, *Governance and Politics in China*, Basingstoke and New York: Palgrave Macmillan 2004; *China Quarterly* Special Issues, 2004.

19 Central Government

Twenty-seven years of reform have been mainly devoted to the creation of a market-led economy. Although this has meant that there is a serious disjuncture between economic and political reforms, the latter, though often unacknowledged, are significant. These include some devolution of power to the provinces, restrictions on tenure in high political office; the streamlining of the bureaucracy through the introduction of measures to improve the quality of its personnel and reducing the number of ministries from 41 to 29; and limiting the state's role in the economy and in the lives of its citizens. The political reforms contribute to a process of stripping away the foundations of a Leninist state, necessary to the development of a constitutional system and the institutionalization of power.

China's constitutional development is likely to be difficult and uneven. Although the end-product of this development cannot be predetermined, the chances are that it will be better-suited and more adaptable to China's needs than a constitution grafted on as a result of a "big bang" approach to constitution-making which assumes that, once drafted and promulgated, the transition to democracy will follow. The 1982 constitution, the fourth since the founding of the People's Republic in 1949, has been amended several times to reflect the market-led economy and, most recently, to strengthen the position of private enterprise, the protection of private property and to incorporate the rule of law.

A pyramidic representation of China's central government provides a clearer picture of the power structures than the constitution affords. Firstly, China's government can be seen to possess those functions commonly associated with a political system: executive, legislative and judicial. Secondly, it is possible to identify those characteristics that are special to China, in particular, the Standing Committee of the National People's Congress (NPC) and the State and Party's Central Military Commissions. Thirdly, while the constitution specifies the National People's Congress as the highest organ of state power it is replaced here by the State Council in respect to the exercise of power. Fourthly, rather than a separation of powers or functions, the branches of government are shown to be linked and in an order according to the power they exercise.

The constitutional designation of the NPC as the highest organ of state power has led commentators to focus on what it does not do rather than what it does do. Too often, it has been dismissed as a rubber stamp legislature that meets once a year to hear the reports of state (and party) leaders. Certainly, the indirect election of its 2,979 members undermines its authority and independence.

Since the mid-1980s, however, the NPC has been attempting to assert its authority. First, its cause has been championed by successive chairmen of the Standing Committee, appointed from the upper echelons of the Party-state power structure during this period. The present chairman, Wu Bangguo, is in a position to further or to curtail this process. Secondly, the economic reforms have been generating a growing volume of legislation, which is being met by a rise in professionalism and institutional resources. Thirdly, there have been changes in legislative behaviour in so far as a sizable number of delegates are beginning to propose amendments to legislation, delay legislation, withhold unanimous approval of appointments and work reports, and introduce their own items for consideration. Fourthly, there is evidence of lobbying mainly by the delegates on behalf of the local interest. Fifthly, the Standing Committee, which meets on a regular basis between the annual sessions of the NPC, is a working body. When the performance of China's incipient legislature is compared with that of more mature legislatures, the development of the NPC stands up reasonably well. When compared with the State Council it is clearly subordinate, but this is not out of line with executive–legislative relations elsewhere.

Sources: Tony Saich, "Reform and the Role of the State in China" in Robert Benewick, et al, (eds) Asian Politics in Development, London and Portland, Oregon: Frank Cass, 2003; Marc Blecher, China Against The Tides, London and New York: Continuum, second edition, 2003; John P. Burns, "Governance and Civil Service Reform", in Jude Howell (ed.), Governance in China, Lanham, Maryland and Oxford: Rowman and Littlefield, 2004; David Shambaugh, (ed.), The Modern Chinese State, New York and Cambridge: Cambridge University Press, 2000.

20 The People's Liberation Army

The state in China is almost always referred to as the Party-state. The structure of the Communist Party parallels that of the government at all levels and crucially dominates the government at each level. An analysis of the Party-state, however, must also include the armed forces, the People's Liberation Army (PLA). This is in part historical, for the PLA was formed in 1927, 22 years before the founding of the People's Republic. It was also formed on the basis of the integration of Party and military leaders. Mao Zedong set out the principle "that the Party commands the gun and the gun must never be allowed to command the Party". The instrument of control is not a state ministry of defense, but the Party's Central Military Commission, which is presided over by the Party General Secretary, Hu Jintao.

The actual relationship between the Party and the PLA is symbiotic and mutually reinforcing. The first two generations of China's leaders derived much of their legitimacy from their participation in the anti-Japanese war and the two civil wars that brought the Communist Party to power. The Party-state relies on the PLA to ensure political stability and to project China as an international power. The PLA has divided loyalties between its historical commitments as a people's army and ambitions to become a modern military force. A modernization program was launched in the mid-1980s to reduce the size of personnel from 4.3 million to 2.25 million.

The Chinese White Paper on National Defense of 1998 stated: "During the new historical period the Chinese Army is working hard to improve its quality and endeavouring to streamline the army the Chinese way, aiming to form a revolutionized, modernized and regularized people's army with Chinese characteristics." This is reflected in the rapid annual growth of defense spending. The Fourth White Paper, published at the end of 2003, confirms the modernization program, while emphasizing "active defense", referring to the reunification of Taiwan, stopping separatist movements and safeguarding political and social stability.

Sources: White Paper on China's National Defense , Information Office, State Council of the People's Republic of China, 1998 • Fourth White Paper on China's National Defense , Information Office, State Council of the People's Republic of China, 2003 • Ray Cheung, "Party Asserts Control over PLA to ease strained ties", South China Morning Post, May 11, 2004 • John Hill, China's armed forces set to undergo face-lift, Jane's Intelligence Review, vol. 26, no. 2, February 2003, pp. 10–16 • Nan Li, "Organizational Changes of the PLA", China Quarterly (no. 158), June 1999 • Colin Mackerras et al (eds.), The Dictionary of The People's Republic of China, London and New York: Routledge, 1998 • David Shambaugh (ed.), The Modern Chinese State, Cambridge and New York: CUP, 2000 • David Shambaugh, Modernizing China's Military, Berkeley and London: University of California Press, 2004 • IISS, The Military Balance, 2004–05, Oxford: Oxford University Press, 2004.

21 The Center and the Provinces

The post-Maoist reforms have produced considerable tensions between the center and the provinces. It is clear, however, that increased local autonomy should not be mistaken for separatism, provincial realignments, or the possibility of federalism (even with Chinese characteristics) in the near future. China is very much a unitary Party-state.

Even so, following fiscal decentralization, there was a shift in power from the center and towards the provinces, and between provinces. The center has attempted to reassert its authority through the introduction of a new tax-sharing system in 1994, which has met with mixed success.

A second feature of local autonomy which gives the provinces a real grip on power is their ability to attract Foreign Direct Investment. This is especially true for those provinces on the eastern seaboard who receive the lion's share.

A third feature has been the growth of extra-budgetary revenue to a size where it almost equals budgetary revenue. The imposition of local taxes and fees, however, is encountering increasing local resistance, forcing the government to once again rethink its taxation policies, including the agricultural tax reform, and experiments in the reforms of personal income taxes, particularly in order to control the levying of fees.

A fourth feature of local autonomy is the development of village governance or micro-states. A number of reforms have resulted in de facto political decentralization from the center. The diagram shows that party and state structures extend in parallel from the center to the grassroots, and that both party and state reproduce themselves at each level. The general principle is that the government at each level is responsible both to the one above and to the party at its own level.

There have been a number of changes since 1978, however. First, standing committees of people's congresses were instituted at the provincial and county levels to supervise, in particular, the work of the government. Secondly, local people's congresses also elect their local government leaders and enact local laws and regulations. Thirdly, direct elections have been extended upwards to the county-level people's congresses. As is the case for the National People's Congress, people's congresses are no longer simply rubber-stamp legislatures. Despite these changes the Party remains firmly in control. The 7,000 senior Party and Government leaders at the provincial level are appointed by Party Organization Department and it is the Party group that exercises real power at all the levels of government.

A dramatic change is the acquisition and exercise of considerable power by the more prosperous villages. Increasingly, they have assumed responsibility for the regulation of land use and the protection of property rights and, as such, behave as micro-states. There have been five rounds of nationwide elections, involving nearly one million villages, since 1988. The elections have been growing more competitive and the use of the secret ballot is not uncommon. Where villagers have suffered, or at least have perceived there to be, abuses of power and mismanagement of resources, they have not hesitated in voting established and party leaders out of office. In addition to elected village committees, representative village assemblies have been created and village constitutions have been drafted.

Village committees exist below the township, or basic, level of government. They are involved in the development of local economies and in the provision of services and welfare. Where the villages are prosperous, the village committees have considerable resources at their disposal, which can lead to intense political debate about their distribution. More recently, cities have been establishing elected community residence committees to improve local services and to increase the delivery, accountability and transparency of the committees.

Sources: Robert Benewick, *et al* "Self-Governance and Community", *China Information*, vol. 8, no. 1, 2004, pp. 11–28 • Robert Benewick, and Akio Takahara, "Eight Grannies with Nine Teeth Between Them", *Journal of Chinese Political Science*, vol. 7 nos 1 & 2, 2002, pp. 1–18 • Benjamin L Reid, "Revitalizing the State's Urban Tips", *The China Quarterly*, Issue 163, September 2001, pp. 806–26 • Akio Takahara, "Managing Central–Local Relations during Socialist Marketization" in Robert Benewick *et al* (eds.) *Politics in Development*, London and Portland, Oregon: Frank Cass, 2003 • Tony Saich, *Governance and Politics of China*, second edition, Basingstoke and New York: Palgrave Macmillan, 2004 • Linda Jacobson, "Local Governance" in Jude Howell (ed.), *Governance in China*, Lanham, Maryland and London, Rowan and Littlefield, 2004 • Zhang Jing, "Neighborhood-Level Governance" in Jude Howell (ed.), *op. cit.*

22 Rule According to Law

The National People's Congress (NPC) at its annual meeting in March 1999 amended the constitution to incorporate the rule of law, although the Party remains supreme. A huge volume of laws are being passed to facilitate foreign trade and investment, and to meet the requirements of WTO membership.

In general, laws in China have less to do with promoting and protecting the rights of citizens than maintaining political and social stability by regulating their behaviour. If we accept that constitutionalism and law is a developmental process, then China has moved from "rule by persons" through "rule by law" on to "rule according to the law".

Corruption is a major obstacle to moving to the stage of the "rule of law". Premier Wen Jiabao has declared that the problem is a matter of "life or death for the Part". Corruption was a major factor leading to the Tiananmen protests of 1989 and it heads the lists of grievances of ordinary Chinese. Popular discontent has been fuelled by accidents caused by poorly constructed roads and bridges. Neglect and oversight threaten the government's ambitious infra-structure programme to stimulate the economy and to create jobs, and also must raise doubts about implementing the rule of law in the near future. There is no shortage of reports that show how endemic and systemic corruption is in China. For example, the Procurator General of the Supreme People's Procuratorate announced that more than 20,000 investigations had been opened on job-related criminal

offences alone during the first six months of 2004 – a 6.9 percent increase over the same period in 2003. At the same time, this demonstrates that the Chinese leadership is taking corruption seriously.

The development of a body of law and of a legal system is accepted as crucial to China's economic modernization but there also must be an acceptance of placing the Party under the rule of law. There has been a rapid growth and professionalization of the legal system. The number of law offices has grown from 8,946 in 1998 to 18,873 in 2002, and similarly the number of lawyers has increased from 101,220 to 136,684 over the same period. That still leaves only 1 lawyer for every 12,700 Chinese, compared with 1 lawyer for every 300 US citizens.

Sources: Robert Benewick, "Towards a Developmental Theory of Constitutionalism: The Chinese Case", *Government and Opposition*, Autumn 1998 • press reports • *China Statistical Yearbook 2004* • Congressional Executive Committee on China, report on Chinese Legal Profession <www.cecc.gov> • Conviction rates: "Executed 'according to the law' – The death penalty in China" <web.amnesty.org/libary> • Human Rights Watch World Report 2003: Asia: China and Tibet <www.hrw.org> • Corruption: a high official of the Central Commission for Discipline Inspection of the CPC, quoted on People's Daily Online <english.peopledaily.com.cn>

23 State *v* Citizens

The title "State *versus* Citizens" signifies how difficult it is for China's leadership to translate its own view of human rights as well as its commitment to international standards into legislative provisions and judicial behaviour. Attention is also drawn to the need for the existence of organizations between citizens and the state, but independent of the state.

The explanation lies not only in China's traditional bias in favor of the state, but also in the paradox of the market. The socialist market-led economy is not, as some would believe, a contradiction in terms. Yet the market does shift the emphasis away from collective and towards individual rights and values, and away from self-sufficiency and towards international inter-dependence and co-operation. The market creates interests, domestically and internationally, and these interests demand and, indeed, need to be heard and expect the state to respond. Market behaviour extends beyond the economy into the formation of citizens' organizations. In China these may be under the leadership of the Communist Party, state-registered organizations, unofficial bodies, underground organizations or political and social movements.

The market, however, also creates new complexities which may threaten state authority and entrenched interests. This new environment of economic openness can engender political volatility. Rather than risk destabilizing conflicts, the Chinese leadership has chosen to promote political stability and the maintenance of public order. In the absence of institutional channels and recognized procedures, such as an independent judiciary and due process of law, the Party-state has resorted to authoritarian means, undermining both human rights and the market values it seeks. The persecution of the relatively small number of political dissidents and intellectuals is one example. The range of crimes punishable by the death penalty and the number of executions that take place is another.

China's trading partners can overlook or dismiss abuses of international standards of human rights in favor of market access and stability. Yet the paradox of the market holds. If China wants international acceptance it will not be able to continue to defy international standards. Although China by no means stands alone as the only state responsible for human rights abuses, as Amnesty reports make clear, it is cast as a pariah among states, particularly when it suits the interests of nations or interests within those nations. This is the case even though it is apparent that reports of the conditions of the labor reform camps (*laogai*), however grim, have been exaggerated, and that political prisoners only make up a small proportion of those incarcerated.

Sources: Amnesty International, *People's Republic of China: Labour Unrest and the Suppression of the Right of Association and Expression*, London: Amnesty International, 2002 • Jude Howell, "New Directions in Civil Society" in Jude Howell (ed.), *Governance in China*, Lanham, Maryland: Rowman & Littlefield, 2004 • Du Jie, "Gender and Governance" in Jude Howell (ed.) (*op.cit.*) • James D. Seymour and Richard Anderson, *New Ghosts, Old Ghosts: Prisons and Labor Reform Camps in China*, Armonk, New York: M. E. Sharpe, 1998 • US Congressional Executive Committee on China, Annual report, 2004 • Gordon White *et al*, *In Search of Civil Society*, London: Macmillan, 1997 • Associated Press, November 23, 2004 • "Paying the Price: Worker Unrest in Northeast China" Human Rights Watch, August 2002 <http://www.hrw.org>

24 Households

The composition of households is demarcated along the rural–urban divide. Rural households are more likely to have extended-family living arrangements, whilst urban households are tending towards single-income households with only two generations living under the same roof. As a result, the problems and advantages of family life are different. Rural families still tend to arrange marriages according to family advantage rather than personal preference. The relationship between mothers-in-law and daughters-in-law is therefore a cogent issue in village life. Competition over household territory and emotional loyalty from son/husband fuels resentment and cruelty between women living in close proximity. In the mid-1980s the Women's Federation (*Fulian*) organized training sessions for young women to learn how to tolerate and accommodate the older woman in their lives. Statistics suggest that living space is not standard between the sectors however. Rural migrants may well live in cramped urban dormitories whilst the urban elites have very comfortable flats. Meanwhile, some farmers are building two-storey housing whilst others are sharing space as their poverty increases.

In major cities extreme poverty is less of an issue, but there is a different concern that wealth in early to middle age might lead to impoverished senior years. Shanghai municipality announced in 2004 that it would take measures to lower the age profile by allowing two-child families for registered residents. The perception that a double income and no kids (DINK) household structure, which made up 12.4 percent of Shanghai households in 2003, will make for a comfortable life does not ease metropolitan fears for an aging population with inadequate pension provisions in place. The privilege is not extended to rural migrants however, who are perceived as a cause of crime and high population density, without quality (*suzhi*).

Meanwhile, domestic violence is an acknowledged but unsolved factor in household relations, with an estimated one in five families experiencing violence. In 2002 a survey by the China Law Society revealed that, "12.1 percent say their husbands kick them, 9.7 percent say their husbands throw things, 5.8 percent say they are forced to have sex and 1.7 percent are burnt or scalded with boiling water". Many rural women are thought to "escape" through suicide (proportionally, there are three times as many suicides in the countryside as in cities – arguably due to the vicious cycle of violence and poverty). In urban areas divorce is on the increase, as is the incidence of single-motherhood in the wake of failed marriages. These social phenomena, perhaps with the exception of high female suicide, are familiar to Westerners, and are accompanied by an equally familiar onus placed on women after the fragmentation of family units.

Sources: Emily Honig and Gail Hershatter, *Personal Voices: Chinese Women in the 1980s*, Stanford University Press, 1988 • Florence Beaugé, "Women's birth right", *Le Monde Diplomatique*, February 1999, p.9. 1988 • "Psychological Domestic Violence Law Proposed" <china.org.cn> translated by Li Liangdu, November 27, 2002 • Antoaneta Bezlova, "Population: Shanghai breaks second-child taboo" International Press Service, September 13, 2004 • "China starts campaigning against domestic violence" ABC Online Correspondents report, June 2004, <www.abc.net.au/correspondents/content/2004/s1135719.htm>

25 Food

The publication of Lester Brown's *Who Will Feed China?* in 1995 fuelled panic at the prospect of a large Chinese population unable to subsist within China's borders. Brown's analysis implied mass migrations and humanitarian catastrophes across Asia by the middle of the 21st century. The rate of shortfall will not be as catastrophic as Brown predicted, but nonetheless, the country is not self-sufficient in basic foodstuffs. China imports soya from Brazil, and grain from the USA, Canada and Australia. In 2004 imports from the USA were 68 percent higher than in 2003. Between 1995 and 2000, the planted area devoted to grains fell from 74 percent to 68 percent, whilst oil-bearing crops rose from 7 percent to 8 percent and vegetables from 6 percent to 10 percent. These shifts are significant markers of the new economy post-WTO and of the tensions between national human food security and trade imperatives.

Differentiated food production is growing in response to the market economy and its effects on population movement, expectations of lifestyle and income, and the emergence of the entrepreneurial state. The shrinkage of fertile land used for grain is in part due to city-bound migration of peasants, but also because of expanding activities in fruit and vegetable production to support farmers' incomes. These products are immediately profitable to China's agricultural industry, although their growth might threaten food security in the longer term. They also serve international and domestic dietary requirements, and, post-WTO entry, encourage foreign direct investment in food processing and branding ventures.

Diet shifts are also crucial in these developments. An increased appetite for meat is leading to an increase in the demand for cheap grain (as animal feed), which may be another factor in choosing to import grain from overseas. There has also been a significant increase in the consumption of fish and other aquatic foods. In 2001, animal husbandry and aquaculture accounted for over 40 percent of "gross agricultural value". The amount of processed food is low in comparison with OECD countries; in 2001 20–25 percent of agricultural produce was processed. The percentage of processed food sold and consumed is rising year by year however, at an estimated rate of 14 percent. Most of this is eaten in urban areas.

The trends in China may be away from a familiar food economy, but the emphasis for such a large, predominantly rural, population must still be the availability and price of grain. An agricultural action plan was announced in March 1999, based on China Agenda 21, which was originally published in 1993, following the 1992 United Nations Conference on Environment and Development. The aim was to achieve sustainable growth in grain production, animal husbandry, fisheries and township enterprises. The first 36 projects included a food-security warning system, water and soil conservation initiatives and animal and plant conservation. The projects calculate that in order to feed a predicted 1.6 billion people in 2030 an annual grain supply of 640 million tons will be needed. The question for China is the degree to which the profits made possible by the WTO should be sought at the expense of grain production and food security on scarce arable land. Although, as a blogger on the subject recently pointed out: "It wouldn't take a great deal of government intervention to get that land back to producing staples."

Sources: Lester R. Brown, *Who Will Feed China?*, New York: W.W. Norton and Co., 1995 • Wen S. Chern, "Projecting Food Demand and Agricultural Trade in China," *The Asia–Pacific Journal of Economics and Business*, vol.1, no. 1, 1997 • EAC (DFAT, Australia), *China Embraces the World Market*, 2003 • <Economymatters.com> • <livingontheplanet.com/asia/> • "China's obesity rate doubles in 10 years years to 60 million people" <health.news.designerz.com> • "China's changing diet" <www.iiasa.ac.at> • "Coca-Cola looks to China for future growth" <www.foodnavigator.com> November 15, 2004

26 Education

In the first decades of the socialist era education was strongly aligned with ideological training, political development, and the formation of good citizens for the new China. This approach to the upbringing of China's youth was in conformity with general concerns in Chinese society and culture. Education (*jiaoyu*) is an essential component of childhood in Chinese life. Children's films, books, outings, must all have a demonstrable educational and patriotic aspect. But in practical terms, giving a working education to China's children is a difficult task. Private education is one of the solutions chosen by parents with the money to pay for it. The Fourteenth Congress of the Chinese Communist Party (1992), made this easier by its endorsement of "the socialist market economy".

Educational inequalities are deepening in the wake of market reforms. Some parents pay out huge sums for private schools and semi-private institutions (*minban*), and all must contribute to their children's education. In 1956 private schools were abolished as part of the reforms of Liberation. By the end of 1993 there were 125 different kinds of private school in Guangdong. Meanwhile, 80 percent of a total of 1 million primary schools have established enterprises through which to better their financial situation. Other children cannot attend school at all when family farming commitments require their labor, or when finances cannot stretch to the cost of fees and books. A developing trend towards streaming exacerbates the disadvantage of children with interrupted or incomplete schooling. Junior high school students (*chuzhong*) are streamed so that some are moved straight into vocational training whilst others continue into senior high school, and consequently enjoy a more extensive set of options. In order to hit nationwide targets for proportional entry from senior high school to university, however, some education regions have significantly reduced the number of senior high places.

As a developing nation China can claim reasonable rates of literacy (88 percent), although they are low when compared to Greater Chinese populations in Hong Kong (91.5 percent) and Taiwan (92 percent). The Education Law attempts to address this through a standard nine years of compulsory schooling – and policy discussions are moving towards a 12-year target. Hu Angang, a senior researcher at Qinghua University, argues for the extended schooling range, pointing out that this will develop the nation's long-term human

resources and skill base, and will also alleviate immediate pressures on employment: "Estimates have stated that by 2005 over 75 percent of middle-school graduates in urban areas will enter high schools, while the proportion in rural areas will reach 65 per cent. By 2010 the percentage in urban areas could increase to almost 100 per cent and the figure in rural areas could reach over 75 per cent. Such an extension (in schooling) will bring China many benefits, primary amongst which will be its role as an effective measure in diminishing strong employment pressure." Hu's theory will only work if those who deliver education policy at the local level (*fangquan*) are able or willing to fund education for all.

The tertiary sector is also growing fast. In 1984 the English language China-Europe Business School (CEIBS), China's first international business school, opened with the support and sponsorship of international firms as well as the local Shanghai Jiaotong University. CEIBS is unusual in having an accredited MBA programme. 4,000 people applied for 120 places in the 1997 admissions round. In 2003, the new Executive MBA students were drawn from all over the world, and if they couldn't afford the fees at CEIBS they might be able to get onto similar programs at Guanghwa School of Management (Beida), or perhaps the Women's Leadership program at Qinghua, all elite and expensive choices. In contrast to students during the Mao era, contemporary students are looking for vocational, technical and commercial relevance in their education. Sichuan University offers courses in real estate, marketing, advertising, and interior decoration. Fudan in Shanghai concentrates on general business skills: accountancy, enterprise management, municipal planning and economics. These courses are self-financing and very popular. Gross enrolment was 12.5 percent in 2000, and in 2002 there were 12 million students enrolled across the tertiary sector. For theorists like Trow and Ma, this represents a shift from elite higher education to mass access, and it signals China's knowing embrace of a knowledge economy.

Sources: Hu Angang, "12 years of schooling will benefit entire nation", CERNET, 2001, *Guangdong Education Committee Yearbook*, 1994 • Mok Ka-ho, "Privatization and Quasi-Marketization: Educational Development in Post-Mao China", Department of Public and Social Administration, City University of Hong Kong, Public and Social Administration Working Paper 1996/4 • "China Survey", *The Economist*, March 8, 1997, pp. 17–19 • Cesar Becani, "Becoming a Foreigner", *Asiaweek*, February 14, 1997, p. 55 • Ka-ho Mok and King-yee Wat, "The Merging of the Public and the Private Boundary: Education and the Market Place in China", CUHK, Working Paper 1997/3 • C. Montgomery Broaded and Chongshun Liu, "Family Background, Gender, and Educational Attainment in Urban China", *The China Quarterly*, 145, March 1996, pp.58–86 • Ma Wanhua, "Higher Education in the Context of Globalisation", Proceedings of the Thirtieth Anniversary International Conference, China Education Centre, 2003

27 Welfare

As the *China Human Development Report* (1998) noted: "Nothing is more important to China's successful transition to a market economy than the completion of a national social insurance program to cope with unemployment, retirement and health needs of the urban population." The Chinese government is acutely aware of this, not only to meet human needs, but to maintain political stability to which it accords the highest priority.

The economic reforms put considerable pressure on the existing welfare services. As the market grows more competitive, state enterprises have found it difficult to carry the burden of their own welfare provisions, including the pensions for their retired workers and support for an over-sized workforce. State-sector employees have also been resistant to losing their employment rights, creating an obstacle to reform. Furthermore, as Gordon White argued, the impact of the economic reforms on the distribution of power has made it difficult for central government to implement a uniform welfare system, even though this has allowed for local flexibility and experimentation. Most important of all have been the demographic changes. The combination of the one-child-family policy and the increase in longevity has had a profound impact on dependency ratios.

Creating a nationwide social insurance system is an immense undertaking. Working in China's favor is its high rate of economic growth, the tradition of family responsibility for the elderly in the countryside, where 61 percent of the population resides, and the decision to delay dismantling the state-owned pillar industries. Yet these factors provide no more than a breathing space, and the difficulties in institutionalizing social welfare outweigh them. The first difficulty is the sheer numbers involved and the need for accurate statistical data. The second is the high cost of such an operation and problem of finding funding. The map on page 75 illustrates the burdens imposed on employed workers by the need to support retired workers.

The third difficulty is that unemployment is likely to be a permanent feature of China's economy. This is compounded by the need to incorporate the mobile population into the welfare system. Fourthly, there is the transition gap. The government has to maintain the existing pensions systems, which are increasingly under funded. Some schemes have to meet the costs for retired workers with limited personal savings because of low wages, while other schemes benefit workers out of proportion to their earnings. Where the new system has been established for state workers, the evidence suggests that the funds have to be used to support current pensioners. One suggestion has been to extend current insurance coverage beyond the state sector. These firms, however, may prefer to have their own plans, rather than subsidize the state sector.

The fifth difficulty is that care has to be taken not to swell the already large numbers of urban poor. This underlines the need for all-encompassing social insurance and for providing an adequate standard of living. Finally, these factors singularly impact on political stability and collectively threaten it.

The objectives of pension reforms have been to shift responsibility away from the enterprises; to share contributions between the workers, enterprises and the state; to include current and future pensioners; and to establish a system that is effective, efficient and fair. The new pension system involves all cities establishing private accounts for every worker into which the worker and the firm pay. They also pay into pooled contingency funds for such eventualities as a firm being unable to meet its pension responsibilities.

Sources: United Nations Development Programme (UNDP), *China Human Development Report*, 1998 • Gordon White, "Social Security Reforms in China: towards an East Asian model" in Roger Goodman *et al* (eds) *The East Asian Welfare Model*, London and New York: Routledge 1998 • Linda Wong, *Marginalization and Social Welfare in China*, London and New York: Routledge 1998 • Richard Jackson and Neil Howe, *The Graying of the Middle Kingdom: The Demographics and Economics of Retirement Policy in China*, Center for Strategic and International Studies, April 2004 <www.csis.org/>; Zeng Yi and Wang Zhenglian, "Dynamics of Family and Elderly Living Arrangements in China: New Lessons Learned from the 2000 Census", *The China Review*, vol. 3, no. 2, Fall 2003 • "The Management of Defence" <www.armedforces.co.uk>

28 Health

The years of famine aside, China under communism has had a good record on health provision. Better results and better services for the majority was the objective of grassroots training and provision. Since the introduction of the market-led economy, the relationship between financial well-being and health has gone into reverse. The symptoms and causes of this decline in provision are manifold. Rural insurance schemes have collapsed due to the demise of collectives and associated co-operative organizations. So-called barefoot doctors no longer receive work points for their service, so now they charge cash. There is a general suspicion of the security of welfare insurance packages, especially in regard to corruption.

Using the most reliable indicator of general health standards, infant mortality, the World Health Organization finds that there was a vast improvement between 1960 and 1985 (173 deaths per 1,000 live births to 44 per 1,000), which improved further to 39 per 1,000 in 2002. However, the indication is that the poorest are still suffering. Women in rural areas have a less than one in two chance of getting a post-natal check-up. In 1999, 77 percent of babies were born at home, 12 percent at a township medical centre, and 9 percent in hospital. Homebirths are not necessarily problematic – indeed, they can help women avoid the complications and infections of hospital doctor-led birthing practices – however, proximity to hospital is a usual condition of homebirth in developed countries, especially on first deliveries. Given the two-child rural policy, most of the births recorded in these figures will be first births. However, the experience of SARS in 2003 and an ongoing Hepatitis B and AIDS crisis, has made the government realize, with the prompting of the World Health Organisation, that the rural population has dire need of extra resources. The China News Service (Xinhua) reported that in 2003 a rural co-operative medicare fund was piloted to reduce the proportion of income required by uninsured rural dwellers to cover healthcare costs. Its success will be premised on the support mechanisms provided by provincial governments, and by nationwide policy initiatives on pandemic prevention. The first breakthrough in this respect was the information campaign launched on World Aids Day, 2003.

China's health system is threatened most directly by the tobacco industry (*pp. 78–79*) and by the spread of HIV/AIDS. Health officials claim that most of the one million mainland Chinese infected by HIV are intravenous drug users. In areas such as Yunnan's border territory with Burma, on the edge of the "Golden Triangle" of Asian drug production, infection rates are estimated at anywhere between 30 and 70 percent . If it is true that drug users are the most vulnerable in China to HIV/AIDS, then the worst-affected area will be in the south. The Health Ministry claims that around 80 percent of users in Guangdong and Guangxi are injecting, compared with 1 percent of users in Shaanxi and Inner Mongolia.

An impending threat to health security comes from the aging population (*see also pp. 74–75*). Senior social economist, Hu Angang has noted that by 2007 those over 64 years will make up 14 percent of the population, and by 2036 this percentage will have increased to 20 percent (at an average rate of 10 million people per year). These people, and society as a whole, will be severely disadvantaged without the immediate introduction of long-term affordable medical insurance and pension schemes.

Sources: Yunniu Wu, "Trends and Opportunities in China's Health Care Sector", Murdoch University: Asia Research Centre Policy Paper, no 18. (1997) • *Green Book of Population and Labour*, CASS Beijing • "Household Consumption and Market Prospects in China",

Murdoch University, Asia Research Centre, 1996 • *China Development Briefing*, Issue 4, January 1997 • October 1996 Speech by Chen Minzhang (Health Minister), in Zhongguo xingbing aizhibing fangzhi zazhi, reported on US Embassy Beijing <http:www.redfish.com> • "Chinese rural dwellers get better medicare service" <www.chinaview.com> October 24, 2004 • "Aging population poses tough challenges for China" <http://english.peopledaily.com.cn> June 9, 2004.

29 Tobacco

It is no secret that tobacco companies in developed countries, faced with a significant decline in smoking, anti-tobacco legislation and lawsuits, anti-tobacco movements and aggressive public hostility towards smokers, are targeting the newly industrialized countries. China beckons, with its huge population, market-oriented economy, impressive economic growth rate, rising living standards, difficulties in enforcing legislation and a thriving tobacco culture in which 70 percent of men smoke.

The irony is that, according to a national survey carried out in 1997, while non-smokers are aware of the risks of smoking, the smokers themselves are either not aware of the anti-smoking arguments or, if addicted, reject them. Over 50 percent of Chinese people believe that smoking does little or no harm, more than 60 percent are unaware of its relation to lung cancer and 96 percent do not know that it can cause heart disease.

Health officials are acutely aware of the risks for China's 360 million smokers and the market inroads that foreign brands are making as fashion icons or by being cheap and available as smuggled items. Their position has been strengthened by the publication in 1998 of the biggest study ever undertaken into deaths from tobacco. Scientists from the USA, the UK, and China investigated 1 million deaths and concluded that if the current smoking uptake rates persist in China, tobacco will kill about 100 million of males currently under the age of 29. Half of these deaths will occur in middle age and half in old age.

Even so, this warning may not be enough to induce government action. As anti-tobacco campaigner Judith Mackay observed in 1997, "the biggest cigarette company in the world is the Chinese government not Philip Morris." China produces more tobacco than any other country in the world and the China National Tobacco Corporation, with about 180 factories, is a state monopoly. Tobacco contributes more than 10 percent to total government revenue and, although health costs may outweigh this, tobacco production is too closely integrated into the state economy for it to be easily unscrambled.

Health officials and anti-tobacco campaigners, for example the China Association on Smoking and Health (CASH), have had their triumphs. A 1995 law banned tobacco advertising on radio, television and in print. Since then, smoking has been forbidden in public places in 82 cities. Enforcement, however, has been uneven; loopholes in the laws have been discovered, and where the bans have clashed with the government's other priorities, the latter have prevailed. For example, Formula One motor racing in China now means that the logos of foreign cigarette brands are attaining an even more glamorous status.

The Party-state's priority – political stability – means that further legislation and effective enforcement is unlikely for the foreseeable future. An increase in taxes may alienate smokers and encourage even more smuggling. Furthermore, the government has no wish to add to unemployment by laying off any of the tobacco processing industry's 230,000 workers, or

disrupting the livelihoods of the 10 million farmers responsible for the 1.3 million hectares of tobacco plants, or of the shopkeepers who rely on the cigarette retail trade.

Sources: Chinese Academy of Preventative Medicine *et al*, *Smoking and Health in China*, 1996 • National Preventative Survey of Smoking Patterns, Beijing: China Science and Technology Press, 1997 • Judith Mackay, *International Herald Tribune*, August 28, 1997 • Liu, Bo-Qi, *et al*, "Emerging Tobacco Hazards in China: 1. Retrospective proportional mortality study of one million deaths," *British Medical Journal*, vol. 317, November 21, 1998 • *Far Eastern Economic Review*, November 26, 1998 • Zhang Jin, "Cigarette sellers cash in on foreign brands" *China Daily*, February 17, 2004 • Global Youth Collaborating Group. Special report: Differences in worldwide tobacco use by gender: findings from the Global Youth Tobacco Survey. *Journal of School Health*, 2003, 73(6): pp. 207–215. Detailed country information available at: <www.cdc.gov/tobacco/global/GYTS.htm>

30 Telecommunications

The concept of globalization is not new. Cross-border trade, migrations and large-scale interdependencies have long ensured that world communities connect. The change in the nature of globalization lies in the collapse of time and space in the wake of advanced systems of communication. Access to the telephone, and in particular the cell phone, together with a developed system of web access, is fundamental to success in 21st-century trade, whether at a high-level policy level or as a street trader.

China has been working in support of a socialist market-led economy to modernize its telecommunications infrastructure systems across the territory. Most provincial capitals are now linked by fiber-optic cabling. The 2005 high-speed train link to Tibet will have digital capacity travelling with it, further enabling China's capacity to connect across its internal borders. Although fixed lines are not adequate – in 2001 only 14 percent of people had a fixed-line connection – cellular phone use and roaming services are expanding very rapidly to accommodate remote users as well as urban customers. Global and local handset manufacturers are competing to supply the market, with new competitors sensitive to the sophistication of software from Linux, Microsoft and the booming games industry in Korea. These developments serve the new economy, but do they promote a different communicative environment?

Internet access in China is provided through educational and government-sponsored establishments. Penetration is still relatively low in comparison with developed countries, but it is high for current levels of income. One in five urban Chinese households owned a computer in 2002. There are also burgeoning internet cafes, although restrictions are periodically imposed to clamp down on inappropriate and politically sensitive sites and searches. Researchers have demonstrated that site blocking can reach 100 percent for certain issues. All of the global top ten sites on Tibet were blocked in one survey, whilst BBC News and the Virtual Islamic University were found to be always unreachable through Chinese servers. Domain-name hijacks are sanctioned by government-sponsored servers, usually resulting in unexpected re-routes through cyber space. Human Rights Watch has drawn attention to the imprisonment of those who defy the content restrictions in university chat-rooms.

The three top providers are CERNET (China Education and Research Network), CSTNET (China Science and Technology Network), and CHINANET (public service). In July 2004 Sina.com – one of China's most successful private internet and mobile providers – announced a new instant messaging service. Two months later it announced that the Chinese regulatory authority had "sanctioned" Sina for exceeding its commercial powers, and required it to suspend its main access number. Given that the Chinese were predicted to send 550 billion short message across 2004 (twice the number sent in 2003), one might pity the cross-media Telco its losses. These events nonetheless suggest the fragility of telecommunications in a highly regulated system. Sina's CEO in 2000 was careful, for instance, to underplay the company's status as a news provider, preferring to emphasize "enabling software" over "content".

Sources: Jonathan Zittrain and Benjamin Edelman, "Empirical Analysis of Internet Filtering in China", <http://cyber.law.harvard.edu> • <http://corp.sina.com.cn> September 26, 2004 • *South China Morning Post*, various articles, 2003–04 • *Asian Communication Handbook*, Singapore: Asian Media Information and Communication Centre, 2001 • Hu Xin, "The Surfer-in-Chief and the Would-be Kings of Content", in Donald, Keane and Yin (eds), *Media in China: Consumption, Content and Crisis*, London: RoutledgeCurzon, 2002 • <http://www.telecomasia.net/> October 24, 2004 • China Embraces the Market, EAU, Department of Foreign Affairs and Trade, Australia • Instat/MDR 2004 • *South China Morning Post*, September 27, 2004

31 Media

The growth of media output and consumption in China is the phenomenal result of a large population, a new-found commitment to consumerism, and the relatively cheap costs of domestic hardware. It is also supported by a centralized state, which can subsidise macro decisions on rollout.

Radio, print media (newspapers, magazines), television (cable, terrestrial and satellite), films (delivered through cinema), video (and compact video-disks), and the internet are the global delivery systems of media product. The question for China, as elsewhere to differing degrees, is the extent to which media systems invite certain types of content. Can media content be controlled and manipulated by audience taste, cultural specificities, and state interests (whether or not these are mutually compatible)? How does the Chinese audience respond to new media choices, and what does the state do to facilitate or block their access?

The internet still poses a challenge to regulation, and the Chinese government has responded with criminal punishments and fines of up to 15,000 yuan (US$1,800) for those who access or provide pornography, anti-Government dissent, political propaganda, and amorality. The list is open-ended and gives the authorities plenty of scope for interpretation according to the priorities of the moment. Regulations are also in place for the monitoring and censorship of other media. One cannot publish photographic essays on the lives or work of national figures, alive or dead. Official biographies are authorized, in print and on film, and are tightly monitored for content according to the political line at the time of release. There are also firm controls on the numbers and titles in the publication and distribution of foreign works, and on the release of foreign movie titles. Journalists comply with a strict code of conduct, but there are newspapers known to be more "mainstream" than others.

Most of the 120 million Chinese households using TV watch local programs, national news bulletins, and "specials". Long-running serials (soap operas) are also extremely popular. Where the content is deemed too boring, especially in the south near Hong Kong and Taiwan, satellites pick up Hong Kong shows. Although in Cantonese, these are more desirable to young putonghua speaking viewers than Hubei music specials. In 1978 China had 32 television stations and 3 million TV sets. Now, there is reportedly 100 percent

penetration and up to 60 free-to-air channels in Chinese households, although that data does not account for villages in remote areas where there is no electricity. Caveats aside, there is significant government commitment to television as a medium, which can be easily monitored – especially through cable delivery. The State Administration for radio, film and television (SARFT) announced in mid-2004 that it planned to facilitate total digitization of television by 2015. The national broadcaster CCTV is now running pay TV channels, and expects to earn more in revenue than it can raise through advertising on its free-to-air services. The pay-TV sector covers movies, music and sport, with some hints that business news may also be provided.

China's film industry has also received a great deal of attention in the last 20 years (since the release of Chen Kaige's *Yellow Earth* in 1984). In recent years, US imports have achieved theatrical release, the quota for which is increased to 20 films per year since WTO accession. *Titanic* (1997), and *Finding Nemo* (2003), were hugely successful. Other titles are available through cheap pirated VCDs (video compact disks) in almost all towns and cities. The effect of pirating also hits local releases, which are often available on the streets before they have been released. The "mainstream melody" (or state sponsored) titles continue to be made for special release on public holidays, but their popularity is limited. The best local success of recent years is the nationalist historical epic *Hero* by Zhang Yimou.

Sources: Michael Keane, "A revolution in television and a great leap forward for innovation? China in the global television format business", in Michael Keane and Albert Moran (eds), *Television across Asia: television industries, programme format and globalisation*, London: RoutledgeCurzon • Anura Goonasekera and Duncan Holaday, *Asian Communication Handbook*, Singapore: Asian Media Information and Communication Centre, 1998 • <http://www.ccnet.com> • Donald, Keane and Yin, *Media in China: Consumption, Content and Crisis*. London: RoutledgeCurzon, 2002 • Jiang Jingjing, "Will nation tune into pay-TV?" *Business Weekly*, August 17, 2004 • "China: Foreign companies to benefit from digital TV rollout" *South China Morning Post*, August 18, 2004 • Yong Zhong, "In search of loyal audiences – what did I find? An ethnographic study of Chinese television audiences", *Continuum: Journal of Media and Cultural Studies*, vol. 17, no.3, 2003, pp. 235–246

32 Religion

China is sometimes seen as a secular state with very little patience for superstition and belief. This is both true and profoundly false. In China there are several ways of understanding religious practice, and several religions to take into collective account when evaluating the state of China's belief. Popular religion, Taoism, Confucianism, Buddhism, Islam, and Christianity are variously present and active across the nation. In all of these it is arguably possible to see the core tenets of Chinese practice being regenerated and pursued: teaching as belief (*jiao*), the worship of Gods and the placation of spirits (both named: *shen*), the recognition of man's place in the cosmos (*tian*). Of course the differences are also profound. Household gods, protection against ghosts, and ancestor worship (*zu xian*) lie at the core of much ritual, although since the founding of the People's Republic in 1949 folk festivals have not been part of the official calendar of national events. Popular Taoism (Daoism) works alongside folk beliefs, local gods and village- or household-centered worship, and is therefore successful and popular. Philosophical Taoism, on the other hand, is concerned with the loss of corporeal signs as motivation in the development of the spirit. Buddhism, and particularly the desire for nirvana, is antithetical to Taoist conceptions of immortality. Christianity and Islam offer a completely different system again, with a focus on a single "true" God, even though each religion disagrees on the nature and commandments of Godliness.

In Taiwan, the majority of religious believers are Buddhists, with a strong influence from Japanese/Chinese schools (Zen/Ch'an). The proportions are: Buddhist 22.8%; Taoist 18.1%, I-Kuan tao 4.4%; Christian 3.4%, other 3.3%. In mainland China the majority of believers are Taoist, there are a substantial number of Buddhists, and a growing number of Muslims and Christians. Many of the Muslims are based in the large minority areas in the west, although there are Muslim communities in eastern cities also. Muslims are susceptible to attack by the State, politically if not in deed, as the Western Provinces become destabilized by the events of September 11, 2001, subsequent wars across the Islamic world, and China's anti-terror alliance with the USA. Christian groups, especially Protestant house churches, are also on the rise, and do not court state approval (thus arguably risking disapproval). Quasi-millennial movements – the most recent being Falun Gong – are seen as an explicit threat to state security, with a leader resident outside China, and with followers who are drawn from those suffering rather than benefiting from the market economy, unemployment, and the end of communist certainties in daily life.

There are other quasi-religious features of Chinese social life, which inflect the basic patterns of belief. Confucian rationality and virtue ethics have been out of fashion since 1949, but it is arguable that the connections between virtue and rationality have continued as a feature of politico-philosophical life. The key words of revolutionary discourse, for example, function similarly to the sacred words of religious texts. As Timothy Cheek, a scholar of the revolution, points out "Confucius held that if names were not correct and realities did not conform to correct names, then the moral state would be an impossibility. The Chinese Communist Party exhibits a faith in the power of names similar to that attributed to Confucius". Their use shapes the meaning of everyday life, by giving it an extra dimension, which must be supported by the belief of speaker and, ideally, of listener. It is the communication of belief that occurs in sacred rites and political rallies that brings the concept of religion close to both that of revolution, and to practical political philosophy.

The bête noire of Chinese religious politics is the Tibetan question. Human rights activists draw on both the right to freedom of religion as well as the desire for autonomy in their defence of Tibetan independence. The Chinese government argues that religion is not a sufficient reason for dividing national territory. There is also a problem where religion and ethnicity come together in a nexus of traditional racisms. The supporters of Tibet have not been so vocal about similar aspirations amongst Muslim Xinjiang separatists. In that case it seems that Western opinion is more in favour of Chinese national stability.

Interestingly, despite the Chinese government's criticism of the figure of the Dalai Lama as a cipher of superstition and feudalism, images of Mao Zedong now appear in village temples and urban fetishism. Whereas before it was the figure and words of Mao that offered a politico-sacred text to his people, now it is his enshrined memory that encapsulates the ancient and the modern in the Chinese religious consciousness. His nirvana is on the earth that made him.

Sources: Joanne O'Brien and Martin Palmer, *The State of Religion Atlas*, New York and London: Simon and Schuster, 1993 • Wang Yiyan, University of Sydney/ SBS lecture notes • Stephan Feuchtwang,

Popular Religion in China: the Imperial Metaphor, London: Curzon, 2001 • Timothy Cheek, "The Names of Rectification: Notes on the Conceptual Domains of CCP Ideology in the Yan'an Rectification Movement" in "Keywords of the Chinese Revolution: The Language of Politics and the Politics of Language in 20th-Century China," funded by the National, Endowment for the Humanities and the Pacific Cultural Foundation <http://www.easc.indiana.edu/>, or hard copy: Indiana University Press • James D. Seymour, "What The Agenda Has Been Missing", in Susan Whitfield (ed.) *After the Event: Human Rights and their Future in China*, London: Wellsweep, 1993 • Stephan Feuchtwang Australia Tibet Council News, December 1998 • *Journal of Chinese Philosophy*, December 1998, vol. 25, nos 1–4 • E. B. Morris, "Philosophic and Religious Content of Chinese Folk Religion," *Chinese Culture: A Quarterly Review* vol. 39, no. 2, June 1998, pp.1–2.8

33 Air Pollution

At the end of 2004 city authorities in Beijing admitted defeat. The target of 227 days (62 percent of the entire year) clear sky was not going to be met, and the "blue skies over Beijing" slogan was under scrutiny. There are many reasons for the failure. Increasing traffic, industrial pollution and sand storms from the Gobi desert are the main offenders. In other cities the story is very similar – indeed, China is the unhappy host to nine of the world's 10 most polluted cities. Air pollution on this scale is a major threat to health in China. Children are described as living in an atmosphere that produces effects equivalent to smoking two packs of cigarettes a day.

Pollution is increased by an over-capacity of fossil-fuel power plants, even in areas expecting to use hydro-electric power in the near future. One explanation for this over capacity is localized short-term productivity gains. The power plants look good, on paper at least, in comparison to other smaller industrial enterprises. They thereby enhance the political reputation and financial standing of local officials. Local plants also work to exclude the state power grid, thus producing more income for local businesses, and higher prices for consumers. Electricity costs fluctuate wildly from province to province and county to county.

A casual regard for air quality results in part from the anarchy in nationwide electricity controls. Coal-smoke is the major pollutant in Chinese cities, causing smog in certain weather conditions. Outside the cities, the effects of fossil-fuel use are exacerbated by a loss of forest cover. Forests in Sichuan and Jiangsu have been decimated since the 1950s, and forest cover along the Yangtze dropped by over 50 percent in the last 30 years of the 20th century.

Rural and urban indoor pollution is also noted by Chinese scientists. The main causes are fossil-fuel stoves (some without flues), and inadequate ventilation, especially in winter. These dangers acutely affect women working at home during the day. High tobacco usage also pollutes living spaces. The government is not unaware of these problems. The Ministry of Agriculture has improved access to safer stoves, and the Beijing City Council attempted to clean up city air by abolishing high-emission taxi-cabs in 1998. Traffic pollution is on a sharp increase and the typical progression from low car volume to increased high-emission traffic is already indicated in the largest conurbations. Sulfur, produced by vehicles using low-grade petrol, is considered by China's State Environment Protection Agency to contribute to 79 percent of the country's smog. There are, however, encouraging signs of Chinese resolve to address the problem, namely a decision by Shanghai City Council to sell only lead-free petrol from the end of 1997, and Beijing's purchase of a fleet of buses running on natural gas.

Sources: *China Daily*, November 4, 2004 (AFP) reported on

<News.Designerz.com> • Song Ruijin, Wang Guifang, Zhou Jinpeng, "Study on the personal exposure level to nitrogen dioxide for housewives in Beijing" in J. J. K. Kaakkola *et al* (eds), *Indoor Air 93: Sixth International Conference on Indoor Air Quality and Climate*, vol. 3, pp. 337–42, Helsinki University of Technology, 1993 • Wang Jin and Y. Zhang , "CO_2 and particle pollution of indoor air in Beijing and its elemental analysis," *Biomedical and Environmental Sciences*, vol. 3, pp. 132–38, 1990 • R. Smith Kirk, Gu Shuhua, Kun Huang, Qiu Daxiong, "One hundred million improved cookstoves in China: How was it done?" *World Development*, vol. 21, no. 6, pp. 941–61, 1993 • *Shanghai Newsletter*, August 9, 1997 <http://www.shanghai-ed.com/> • Sulfur figure quoted in Jonathan Watts, "Toxic smog shrouds Beijing", <http://www.guardian.co.uk> October 11, 2004 • "New "green" engines for Beijing's buses", *Beijing Portal*, 21 April, 2004 <www.beijingportal.com.cn> • Jonathan Fenby, *Observer*, August 18, 2004 • State Environmental Protection Administration (SEPA), cited in *China Daily,* July 15, 2004, quoted in <http://www.china.org.cn > • Scott Hillis, Reuters, "China city fights environmental trouble with trees" quoted on SEPA website <www.zhb.gov.cn>

34 Water Resources

China's water is unevenly distributed, and some would argue that climate change is making weather patterns less predictable. Seasonal flooding appears to be getting more extreme, and droughts more frequent and more severe. The summer of 2004 saw both disasters strike simultaneously. In eastern and central China droughts left around 5 million rural people and 2.7 million head of livestock short of drinking water. Some areas were first affected by drought, then devastated by floodwaters and mud-slides. The common aspect is a flooded south and a dry north. Desertification caused by bad land management is a particular problem, contributing to air pollution as well as water seepage and waste.

The monumental Three Gorges dam was developed to deal with flooding problems, facilitate water-borne transport systems, and provide hydro-electric power, but has been hugely controversial. Questions have been raised about the migration and resettlement of local populations, the quality of the project design, the eventual cost of power, but above all experts have questioned whether it will be effective in controlling the vast power of the Yangtze. The dam is a byword for the water conflicts that are reported (and rumoured) to take place in China, and which also simmer at its borders. Dam construction dislocates vast communities and local activists contest new dams and new disruption; upstream polluters cause tension for downstream users; and – on the international scale – China can access, and pollute, the origins of its neighbours' water sources (such as the Mekong River). All of these problems need to be addressed by inclusive strategies for management, grievance resolution systems, and relevant technology.

In the plains of the northeast, agriculture and industry make heavy demands on available surface water, to the extent that the Yellow river regularly runs dry before reaching the sea. Irrigation is using increasing amounts of water (an 18 percent increase from 1990 to 2002), and critics point to the excessive wastage involved. An increasing reliance is being placed on underground water, which is being used at an unsustainable rate. In some places the water table is dropping by 3 meters a year.

An ambitious "solution" has been devised: to channel water from the south to the north. Although this major engineering feat will, when completed in 2050, result in the diversion of a flow of water similar to that of the Rhine, in Germany, it will still only meet a small proportion of the north's water deficit. In addition, the water will be used, legitimately and illegally, along the way; there is expected to be a high rate of evaporation, and levels of pollutants will increase during the transportation.

Sources: Jane Sayers, Environment Victoria; Xinhuanet, July 25, 2004 • *Shenzhen Daily*, July 28, 2004 <www.sznews.com/szdaily>• Payal Sampat, "Groundwater Mining", WorldWatch Institute, 2000 • "South-to-North Water Diversion Project, China" <www.water-technology.net> • Woodrow Wilson China Environment Series (2005, forthcoming) • Reports on ReliefWeb, including those from: Agence France-Presse, Dec 15, 2003; Action by Churches Together International, July 20, 2004; Deutsche Presse Agentur, July 20, 2004; Reuters, July 22, 2004; IFRC, Sept 10, 2004; IFRC, October 28, 2004 • Report on China Economic Net, August 6, 2004 • Report on Xinhuanet, 12 August 2004

35 Water Health

China's water use – two-thirds for agriculture, and just over one tenth for domestic purposes – is in line with the way water is used in the rest of the world. Domestic use does, however, vary widely between provinces, and between rural and urban areas. In most provinces urban residents use significantly more water in their home than do their rural neighbours. This is probably because access to tap water is more common in urban areas than rural, but lack of information on access to tap water among rural households makes this difficult to verify. Government statistics continue to trumpet the steadily improving access among urban households. Such claims are complicated by the reality of shared households, in old blocks, where people living on higher storeys do not have taps on their floors, in which case "access" is actually severely curtailed in practice.

A WHO/UNICEF report on the proportion of households with access to "improved water sources" (a term that includes communal covered wells) gives estimates based on a number of different sources of 94 percent in urban areas and 66 percent in rural. The estimate for access to adequate sanitation facilities (latrines or toilets that afford privacy, and prevent contact with human waste by insects and animals) are 69 percent for urban and 27 percent for rural.

The level of contaminants in industrial and domestic waste-water has an impact on the environment in general, and on the quality of drinking water. The emission of untreated or poorly treated domestic wastewater is increasing. Polluted drinking water is also caused by the overuse of chemical fertilizers in agriculture, and large-scale garbage dumping. China has the highest global rate of organic pollutants, and the entry of heavy metals into the food chain from irrigation and seepage is concerning for long-term health. Industrial water pollution has been targeted for improvement, and domestic reports claim some success, but that is disputed by environmental NGOs in China, who point to the lack of resources for implementing anti-pollution statutes.

Sources: WHO/UNICEF Joint Monitoring Programme for Water Supply and Sanitation Coverage Estimates 1980–2000. *Access to Improved Drinking Water Sources: China*, September 2001 • China's water crisis blamed on waste and pollution, Xinhua News Agency, July 21, 1998 • G. K. Heilig, *RAPS-China. A Regional Analysis and Planning System*, Laxenburg, Austria, 2004

Part 7: Tables

"First Impressions of the 2000 Census of China", November 4, 2001, William Lavely (Sociology Department and Center for Studies in Demography and Ecology University of Washington), published on internet

⊕ SELECT BIBLIOGRAPHY

Primary sources

China Daily
  <www.chinadaily.com.cn/english/home/index.html>
China Labour Statistical Yearbook, Beijing: China Statistical
  Publishing House, latest year
China Population Statistics Yearbook, Beijing: China
  Statistical Publishing House, latest year
China Quarterly various issues
China Statistical Yearbook, Beijing: China Statistical
  Publishing House, latest year
Far Eastern Economic Review various issues
International Institute of Strategic Studies (IISS), *The Military
  Balance, 2004–2005*, Oxford: Oxford University Press, 2004
People's Daily Online
United Nations Development Programme (UNDP), *China
  Human Development Report*, United Nations, 1998
United Nations Development Programme (UNDP), *Human
  Development Report*, latest year
US Census Bureau, International Database
  <http://www.census.gov/ipc/www/idbnew.html>
World Bank, *The World Development Indicators*, Washington:
  World Bank, annual publication
Xinhua.Net <http://www.xinhuanet.com/english/>

Secondary sources

Benewick, Robert et al (eds), *Asian Politics in Development*,
  London & Portland, Oregon: Frank Cass, 2003
Blecher, Marc, *China Against the Tides*, New York: Continuum,
  2003
Buzan, Barry & Foot, Rosemary, *Does China Matter?*, London
  & New York: RoutledgeCurzon 2004
Chih-yu Shih: *Negotiating Ethnicity in China: Citizenship as a
  response to the state*, Routledge 2002
China Environment Series, Woodrow Wilson Center,
  Washington DC
Department of Foreign Affairs and Trade, Australia: "China
  Embraces the World Market" <dfat.gov.au/eau> 2002
Donald, S.H., Keane, Michael & Yin Hong, *Media in China:
  Consumption, Content and crisis*, London:
  RoutledgeCurzon, 2002
Duckett, J., *The Entrepreneurial State in China*, London:
  Routledge, 1998
Fang Shan, "Unemployment in Mainland China: Current
  situation and possible trends", *Issues and Studies*, 32–10,
  October 1996
Foot, Rosemary, *Rights Beyond Borders*, Oxford: Oxford
  University Press, 2000
Goodman, David S.G. (ed.), *China's Provinces in Reform*,
  London and New York: Routledge, 1997
Goodman, David S.G. "Why Women Count: Chinese women
  and the leadership of reform", in Anne E. Maclaren (ed.),
  Chinese Women: Living and working, London:
  RoutledgeCurzon, 2004
Goodman, David S.G., "Qinghai and the Emergence of the
  West: Nationalities, communal interaction, and national
  integration", *The China Quarterly*, no 178, June 2004
Hamilton, Gary G. (ed.), *Cosmopolitan Capitalists: Hong Kong
  and the Chinese Diaspora at the end of the twentieth
  century*, Seattle: University of Washington Press, 1999

Harrell, Steven (ed.), *Cultural Encounters on China's Ethnic
  Frontiers*, Seattle: University of Washington Press, 1995
Holbig, Heike & Robert Ash, *China's Accession to the World
  Trade Organization: National and international
  Perspectives*, London: RoutledgeCurzon, 2002
Howell, Jude (ed.) *Governance in China,* Lanham, Maryland:
  Roman and Littlefield, 2004
Jacka, Tamara, *Women's Work in China: Change and
  continuity in an era of reform*, Cambridge and Melbourne:
  Cambridge University Press, 1997
Krug, Barbara (ed.), *China's Rational Entrepreneurs: The
  development of the new private business sector*, London:
  RoutledgeCurzon, 2004
Lardy, Nicholas R., *Integrating China into the Global
  Economy*, Washington: Brookings Institution Press, 2002
Li, F. M. Sarah & Sofield, H. B. Trevor, "Tourism Development
  and Socio-cultural Change in Rural China", A. V. Seaton
  (ed.), *Tourism: The state of the srt*, Chichester: John Wiley
  & Sons, 1994
Lipman, Jonathan N: *Familiar Strangers: A history of
  Muslims in northwest China*, Washington: University of
  Washington Press, 1997
Mackay, Judith & Eriksen, Michael, *The Tobacco Atlas*,
  Geneva: World Health Organization, 2002
Mackerras, Colin, *China's Minorities*, Hong Kong: Oxford
  University Press, 1994
Mackerras, Colin, *et al*, (eds), *Dictionary of the Politics of the
  People's Republic of China*, London and New York:
  Routledge, 1998
Marton, Andrew M., *China's Spatial Economic Development:
  Regional yransformation in the Lower Yangzi Delta*,
  London: Routledge, 2000
Moran, Albert & Keane, Michael (eds), *Television across Asia:
  Television industries, programme formats and
  globalization*, London: RoutledgeCurzon 2004
Nolan, Peter, *China at the Crossroads*, Cambridge: Polity
  Press, 2004
O'Brien, Joanne and Martin Palmer, *The State of Religion
  Atlas*, New York and London: Simon and Schuster, 1993
Oakes, Tim, *Tourism and Modernity in China,* London:
  RoutledgeCurzon, 1998
Saich, Tony, *Governance and Politics of China*, Basingstoke
  and New York: Palgrave Macmillan, 2004
Seymour, James D. and Anderson, Richard, *New Ghosts, Old
  Ghosts: Prisons and labor reform camps in China*, Armonk,
  New York: M. E. Sharpe, 1998
Shambaugh, David (ed.), *The Modern Chinese State*,
  Cambridge: Cambridge University Press, 2000
Shambaugh, David, *Modernizing China's Military*, Berkeley
  and London: University of California Press, 2004
Smith, Dan, *The Atlas of War and Peace Atlas*, London:
  Earthscan; New York: Penguin, 2003
White, Gordon, et al, *In Search of Civil Society*, London:
  Macmillan, 1997
Yuezhi Zhao, *Media market, and Democracy in China:
  Between the party line and the bottom line*, Chicago:
  University of Illinois Press, 1998
Zweig, David, *Internationalizing China: Domestic Interests
  and Global Linkages*, Cornell University Press, 2002

accidents 118
   industrial 114
   road traffic 48–49, 115, 118
aging population 74, 75, 121
agricultural labor 23
agriculture 33, 40–41, 101, 113, 119,
   122, 125
air pollution 48, 87, 88–89, 124–25,
   125
   indoor 124
air travel 48, 115
All-China Federation of Trade
   Unions 65
All-China Women's Federation 65
appeals system 62
armed forces 19, 58
Asia–Pacific Economic
   Cooperation forum (APEC) 112
Association of South East Asian
   Nations (ASEAN) 21, 112
Australia 115, 119
automobiles see cars
Autonomous Regions (ARs) 30

Beijing 46, 67, 115, 124
books 82
Buddhism, 84, 85, 111, 123–24

carbon dioxide emissions 88
cars 33, 48–49, 102, 115
cash crops 113–12
cell phones 80, 103, 122
censorship 64, 122–23, 123
Central Military Commission 54,
   57, 58, 116
Chinese Communist Party 13, 53,
   54–55, 56–57, 60, 116–19
Chinese Women's National
   Congress 111
Christianity 67, 84, 123–24 see also
   house churches
cigarettes 78, 79
cinema see film industry
class structure 38
Closer Economic Partnership
   Arrangement 16, 112
coal 50, 51, 105
   mines 114
Communist Youth League of China
   65
Community Residents'
   Committees 53, 68
compact discs 83
Confucianism 123–24

conservation 120
constitution 62, 63
consumer goods 37
contraception 28, 95, 110
copyright see intellectual property
   rights
corruption 62, 114, 118, 121
crop yield 41

Dalai Lama 85, 124
death
   causes of 77
   penalty 64
deforestation 87, 124
democratic parties 54
Deng Xiaoping 13, 34, 95, 116
dependency ratio 75, 121
desertification 87, 125
diet 70, 76
divorce 68
domestic violence 119
drought 41, 87, 90, 125

East Asian Economic Caucus
   (EAEC) 112
economy
   growth of 13, 14, 16, 33, 87, 108,
      122
   liberalization of 14, 16, 53, 64, 95,
      116, 119, 122
education 72–73, 112, 120
elections 53, 117, 118
electricity 50, 51, 105, 115
employment 38–39, 40, 42, 44, 97,
   102
   private 38
energy 50–51, 87, 105
entrepreneurs 33, 34–35, 53, 54,
   113, 114
environment 33, 87, 88–93, 108, 111
European Union (EU) 20
extra-budgetary revenue 117

factory farming 40
Falun Gong 64, 67, 84, 124
fast food 33, 67, 70
female trafficking 28, 29
film industry 67, 82, 123
financial markets 16–17, 108, 114
fishing 41, 71, 119
floods 87, 90, 125
food 67, 70–71, 102, 119–20
Foreign Direct Investment 16–17,
   43, 61, 97, 100, 108, 117, 119

gas
   industrial emissions 88, 105
   natural 51, 106, 115
gender disparity 23, 28–29, 65, 98,
   110, 112
Go West program 87, 108
government 56–57, 61, 108, 116,
   117–18
   policy 13
   revenue 60
grain production 33, 40, 41, 108,
   113, 119, 120
Greater China 112
gross domestic product (GDP) 37,
   40, 42, 44, 45, 96, 100
guanxi 62, 114

health 76–77, 103, 121, 125
HIV/AIDS 64, 76, 121
Hong Kong SAR 15, 16, 56, 112,
   114, 115, 120, 123
horizontal regional cooperation
   112
house churches 65, 84, 124
households 68–69, 119
Hu Jintao 53, 55, 58, 116, 117
human rights 13, 20, 64, 65, 108,
   118, 124
hydro-power 50

illiteracy 37, 73, 99, 112, 120
   among women 73, 111
India 21
industrial gas emissions 88, 89
industry 42–43, 89, 101, 108, 109
inequality 33, 36–37, 67, 76, 108, 112
   rural–urban 23, 36, 37, 72, 93,
      102, 112, 119, 121, 125
infant mortality 121
infanticide 28
information technology 44, 45, 114
intellectual property rights 16, 20
intellectuals 54
international relations 13, 20–21,
   109
internet 81, 103, 122, 123
irrigation 40, 125
Islam 67, 111, 123–24

Japan 15, 20, 21, 108, 109, 113
Jiang Zemin 53, 116

Kazakhstan 21, 111
Kyoto Protocol 115

Kyrgyzstan 21, 111

labor 13, 108, 109
law 62–63, 118
   expenditure on 63
   rule of 116, 118
life expectancy 96, 98, 111, 121
living arrangements 68, 75

Macau SAR, 112
magazines 67, 82
Malaysia 20
Mao Zedong 95, 117, 124
market economy *see* economy
media 67, 82–83, 123
mediation 62, 63
migrant workers 23, 26, 67, 95, 110,
   111, 113, 114, 119
military 13, 16–17, 53, 57, 58–59,
   108–9, 116, 117
   expenditure on 18, 59, 108
minority nationalities 30–31, 53,
   111, 112
mobile phones *see* cell phones
motorized transport 48–49, 115

National People's Congress 56, 111,
   116, 117, 118
newspapers 67, 82
non-government organizations 53,
   65
North Korea 21
nuclear power 50, 116
nuclear weapons 19

obesity 70
oil 21, 50, 51, 105, 108, 115
Olympics, 2008, 13, 46, 64, 89
one-child-family policy 28, 29, 67,
   110, 121
opening-up *see* trade
overseas students 72

Party Leading Groups 116
pensions 67, 74, 109, 120–21, 121
People's Liberation Army *see*
   military
Philippines 20, 21
pillar organizations 53, 65
Politburo 54, 55, 57 116–17
political power 54–55, 56–57,
   60–61,
   devolution of 108, 116, 117–18
   distance from 112, 117–18

pollution 33, 76, 87
   air 48, 87, 88–89, 124–25
   chemical 87, 113, 125
population 23, 24–25, 87, 96, 98,
   109, 110
   density 24, 96
   dependency ratio 75, 99, 109
   growth 24, 96, 98, 109
   minority nationalities 27, 99, 111,
      112
   rural 27, 110
   urban 27, 99
private sector 35, 38, 43
   enterprise 33, 34–35, 53, 54, 113,
      114, 116
   education 73, 120
   housing 35, 68, 69, 116
   savings 34

radio 82
rail transport 48, 115
real-estate development 35, 68,
   112, 114, 120
registration system
   for individuals 26
   for organizations 53, 64
religion 67, 84–85, 111, 123–24
religious repression 111
Russia 21, 109

sanitation 67, 92
SARS 77, 121
services 44–45, 101, 113, 114, 120
Shanghai 33, 46, 115, 119, 124
Shanghai Cooperation
   Organization 13, 20
smoking *see* tobacco
South America 21
South China Sea 20
South Korea 21, 109
Standing Committee of the
   Politburo 55, 57, 116–17
State Environmental Protection
   Agency 88
state-owned enterprises 42, 43,
   114, 120–21
sterilization 28, 67
sulfur 88, 124

Taiwan 13, 20, 21, 59, 58, 108, 109,
   112, 120, 123, 124
Tajikistan 21, 111
Taoism 84, 123–24
telecommunications 80–81, 108,
   122–23
telephone subscribers 81, 112, 122

television 67, 82, 123
   ownership of 37, 83
terrorism 13, 109, 124
Three Gorges Dam 90, 125
Three Represents 64
Tiananmen Square 20, 116
Tibet 30, 111, 115, 124
   fast rail link 115, 122
tobacco 76, 78–79, 103, 122
tourism 33, 34, 46–47, 100, 112,
   114–15
trade 13, 14–15, 20, 21, 23, 42, 97,
   108, 109
trade unions 64
traffic 48–49, 115, 124
Turkmenistan 111

under-five mortality rate 76
underground organizations 53, 65
unemployment 39, 95, 108, 109,
   113, 121

United Nations 13, 20, 119
   peacekeeping operations 18, 109
urban employment 39, 102
urbanization 26–27, 110, 113
USA 13, 14, 21, 108, 109, 119, 124
Uygurs 13, 30, 31, 111
Uzbekistan 21, 111

video compact discs 67, 83
Vietnam 20, 21
village committees 118

wages 39, 102
water 87, 90–91, 92–93, 95, 104, 125
   pollution 92, 104, 125
welfare 33, 74–75, 95, 112, 114,
   120–21
Wen Jiabao 116, 118
Western Provinces 33, 112, 115, 124
women
   entrepreneurs 34, 110
   and contraception 28, 110
   and employment, 39 111, 113
   in public life 111
   in trade unions 65
World Health Organization 121
World Trade Organization 14, 16,
   33, 78, 108, 109, 112, 113, 118,
   119, 120, 123
Wu Bangguo 116, 117

Yangtze 90, 91 125